SOLID GEOMETRY

BY

P. M. COHN

ROUTLEDGE AND KEGAN PAUL
LONDON

First published 1961
Routledge & Kegan Paul Ltd
Broadway House, 68–74 Carter Lane
London, E.C.4

Second impression 1963
Third impression 1965
Fourth impression 1968

Printed in Great Britain by
Latimer Trend & Co. Ltd., Plymouth

SBN 7100 6343 1

Preface

Many problems in science and engineering require some knowledge of three-dimensional geometry, and the best way to acquire such knowledge today is by using vectors and matrices. This method has the advantage that the geometry can then be used to illustrate the algebraic concepts, so that the student can consolidate and extend his knowledge of vector and matrix-theory. The object of this book is to discuss lines, planes, spheres and central quadrics in this way. There is also a chapter on coordinate transformations which are essential to a study of quadrics and which provide a good application of matrix-theory.

The definition and elementary properties of matrices which are needed here may be obtained from the author's *Linear Equations* (also in this series) or any book on linear algebra. On the other hand, an account of vectors from the geometric point of view has been included; this is to complement the algebraic treatment in *Linear Equations* and also to make the earlier chapters independent of that text.

It is again a pleasant duty to thank Dr. W. Ledermann for his helpful suggestions.

P. M. COHN

The University,
Manchester.

Contents

CHAPTER ONE

Vectors and Coordinate-Systems

1. In solid geometry we study the relation of points in space to each other. To do this we need a means of passing from one point to another, and the simplest such means is a *translation*. This may be thought of as an instruction to move a certain distance in a certain direction, independently of the starting point; thus if every one of a group of people moves two yards north, their relative positions are unchanged and we have a translation. Given any two points P and P' in space, there is just one translation which moves an object situated at P to P'; this translation displaces any object in the direction of the line PP' by an amount equal to the distance PP'. If the same translation moves an object situated at another point Q to Q', then it may also be defined in terms of Q and Q' (Fig. 1); thus the translation

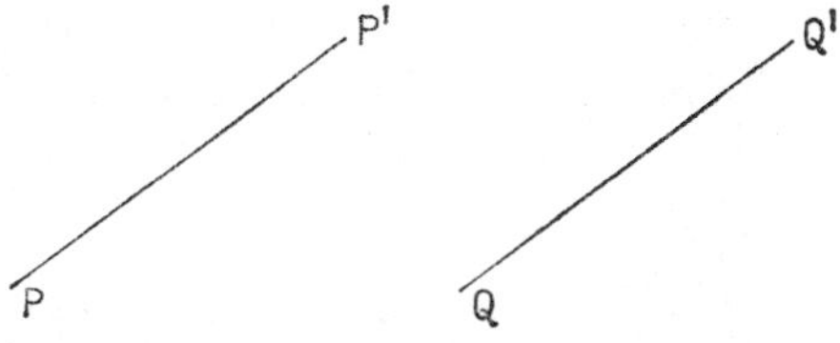

FIG. 1

depends, not on P itself, but only on the length and direction of the segment PP'. It is convenient to have a separate name for this notion of 'length and direction', which serves to specify a translation. We call it a *vector*† and usually

† The word 'vector', introduced by Sir W. R. Hamilton (1805–65), means 'carrier', expressing the fact that a vector moves or carries each (geometrical) object a fixed distance in a fixed direction.

denote it by a lower-case letter in bold-face type.

We can sum up this description of vectors by saying that:

(i) A vector **a** assigns to each point P of space a point P' such that the length and direction of the segment PP' is the same for each point P.

(ii) Any two points P and P' of space define a vector, namely the vector which assigns P' to P.

The vector described in (ii) is often denoted by PP'; in this notation the vector **a** in (i) may be written PP', or equally well, QQ', where Q is any other point, and Q' the point assigned to it by **a**.

The concept of vector laid down in (i) and (ii) is basic in all that follows. Often the vectors will be represented by line-segments, and since a line-segment to represent a given vector may be drawn from any point, the line-segments representing the different vectors are drawn from a fixed point O say, which is called the *origin*. The vector OP is then called *the position-vector of P relative to O*, or simply *the vector of P relative to O*.

2. If we perform two translations in succession we again obtain a translation. Intuitively this is clear if we think of translations as moving a large body like a solid block, without turning. On the other hand, it can also be shown geometrically that in applying two translations in succession, everything is displaced by the same amount in the same direction, though we shall not do so here.†

Let **a** and **b** be any vectors. If the translations corresponding to **a** and **b** are carried out (in that order), the result is a translation whose vector we denote by $\mathbf{a}+\mathbf{b}$ and call the *sum* of **a** and **b**. If **a** is represented by PQ and **b** by QR, then $\mathbf{a}+\mathbf{b}$ moves from P to R, and hence may be represented by PR. This is often expressed as the *parallelogram rule* of vector addition (Fig. 2):

† This amounts to showing that two triangles which agree in two sides and the included angle, are congruent.

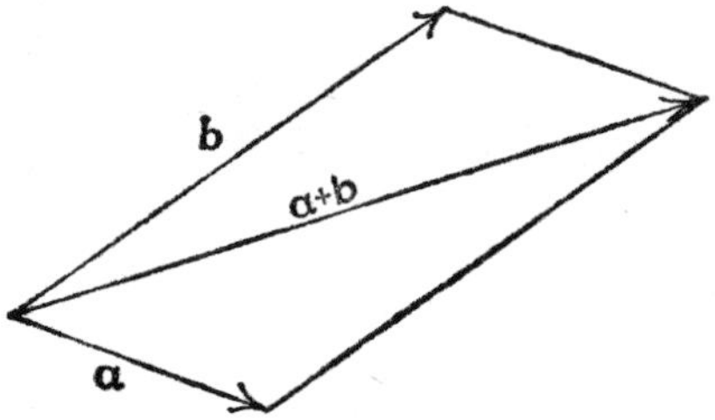

FIG. 2

If two vectors **a**, **b** are represented by adjacent sides of a parallelogram, then their sum is represented by the diagonal joining their extremities.

In applying this rule it is important to distinguish a vector such as PQ from the vector QP which points in the opposite direction. The vector QP is also written $-PQ$; generally if **a** is the vector which goes from P to Q, then the vector which goes from Q to P is denoted by $-\mathbf{a}$.

A particular translation is that in which everything stays where it is. The corresponding vector is called the *zero-vector* and is denoted by **o**.

The laws of vector addition may now be stated:

V.1 $(\mathbf{a}+\mathbf{b})+\mathbf{c}=\mathbf{a}+(\mathbf{b}+\mathbf{c})$ (associative law),

V.2. $\mathbf{a}+\mathbf{b}=\mathbf{b}+\mathbf{a}$ (commutative law),

V.3. $\mathbf{a}+\mathbf{o}=\mathbf{a}$,

V.4. $\mathbf{a}+(-\mathbf{a})=\mathbf{o}$.

These laws may be verified without much difficulty, using the definitions and some elementary geometrical facts. Thus V.1 follows because both sides represent the vector of the translation obtained by carrying out the translations corresponding to the vectors **a**, **b**, **c**, in that order (Fig. 3). V.2 follows from the theorem that if in a quadrilateral two opposite sides are equal and parallel, then so are the other two. Finally V.3 and V.4 are immediate consequences of the definitions of **o** and of $-\mathbf{a}$.

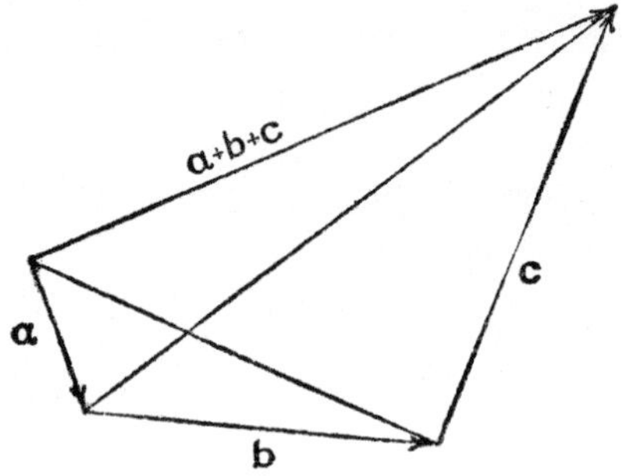

FIG. 3

3. From the associative and commutative laws (V.1 and V.2) it follows that in forming sums of vectors we need not put brackets or pay attention to the order of the vectors, just as in adding numbers. In particular, we can form multiples of a given vector

$$n\mathbf{a}=\mathbf{a}+\mathbf{a}+\ldots+\mathbf{a} \quad (n \text{ terms}). \qquad (1)$$

Let **a** be a given vector, not the zero-vector. By applying translations corresponding to the vectors **a**, 2**a**, 3**a**, ... at a point P we reach a series of points P', P'', P''', ... all on a straight line through P and at equal distances. If we want to reach *all* the points on this line we must introduce multiples of vectors by arbitrary real numbers.

Thus for any positive real number λ, $\lambda\mathbf{a}$ represents a vector in the same direction as **a**, but λ times as long; when λ is negative, $\lambda\mathbf{a}$ is in the direction opposite to **a**, and† $|\lambda|$ times as long, while $0\mathbf{a}=\mathbf{o}$. In contrast to vectors, numbers are often called *scalars*.§

The operation just described of multiplying a vector by a scalar satisfies the following laws:

† The absolute value, $|\lambda|$, of a real number λ is defined as λ if $\lambda \geqslant 0$, and $-\lambda$, if $\lambda < 0$.

§ Because they describe quantities which can be read off on a scale.

V.5 $\lambda(\mathbf{a}+\mathbf{b})=\lambda\mathbf{a}+\lambda\mathbf{b}$,
V.6. $(\lambda+\mu)\mathbf{a}=\lambda\mathbf{a}+\mu\mathbf{a}$,
V.7. $(\lambda\mu)\mathbf{a}=\lambda(\mu\mathbf{a})$,
V.8. $1\mathbf{a}=\mathbf{a}$,
V.9. $0\mathbf{a}=\mathbf{o}$.

The task of verifying these laws may again be left to the reader. Thus V.5 expresses a proposition about similar triangles, while V.6–9 are direct consequences of the definition.

4. From a given set of vectors $\mathbf{a}_1, \ldots, \mathbf{a}_r$ we can form others by addition and multiplication by scalars. It is not hard to see, using V.1–9, that the most general vector which can be formed in this way is of the form

$$\lambda_1\mathbf{a}_1+\ldots+\lambda_r\mathbf{a}_r, \tag{2}$$

where $\lambda_1, \ldots, \lambda_r$ are scalars. A vector $\mathbf{b}$ of the form (2) is called a *linear combination* of $\mathbf{a}_1, \ldots, \mathbf{a}_r$; we also say: $\mathbf{b}$ is *linearly dependent* on $\mathbf{a}_1, \ldots, \mathbf{a}_r$. Thus e.g. $\frac{3}{4}\mathbf{a}$ is a linear combination of $\mathbf{a}$, and the zero-vector is linearly dependent on any set of vectors.

A set of vectors $\mathbf{a}_1, \ldots, \mathbf{a}_r$ is said to be *linearly dependent* if one of them is linearly dependent on the rest, otherwise it is *linearly independent*. An equivalent form of this definition is as follows:†

The vectors $\mathbf{a}_1, \ldots, \mathbf{a}_r$ are linearly dependent if and only if there is a relation

$$\alpha_1\mathbf{a}_1+\ldots+\alpha_r\mathbf{a}_r=\mathbf{o},$$

which is *non-trivial*, i.e. in which not all the α's are zero.

Intuitively speaking we can say that two vectors are linearly dependent if they have the same direction (or if at least one of them is $\mathbf{o}$), i.e. if they lie along the same straight line. Three vectors are linearly dependent if they lie in the same plane. The limitation of space to three dimensions is now expressed by the following condition:

† This is proved, e.g. in the author's *Linear Equations*, also in this series. This book will henceforth be referred to as LE.

There exist sets of three linearly independent vectors, but any four vectors are linearly dependent.

5. Let $\mathbf{e}_1$, $\mathbf{e}_2$, $\mathbf{e}_3$ be a given set of three linearly independent vectors, and let $\mathbf{a}$ be any vector. Then the four vectors $\mathbf{a}$, $\mathbf{e}_1$, $\mathbf{e}_2$, $\mathbf{e}_3$ are linearly dependent, i.e.

$$\alpha\mathbf{a}+\alpha_1\mathbf{e}_1+\alpha_2\mathbf{e}_2+\alpha_3\mathbf{e}_3=\mathbf{o}, \qquad (3)$$

where α, α_1, α_2, α_3 are not all zero. In particular, α cannot be zero, because otherwise we would have a non-trivial relation between $\mathbf{e}_1$, $\mathbf{e}_2$, $\mathbf{e}_3$, which by hypothesis cannot exist. Therefore $\alpha\neq 0$; dividing by α and rearranging the terms we can write (3) as

$$\mathbf{a}=a_1\mathbf{e}_1+a_2\mathbf{e}_2+a_3\mathbf{e}_3 \quad (a_i=-\alpha_i/\alpha;\ i=1, 2, 3). \qquad (4)$$

In equation (4) the scalars a_1, a_2, a_3 are uniquely determined by $\mathbf{a}$, for if we had also

$$\mathbf{a}=b_1\mathbf{e}_1+b_2\mathbf{e}_2+b_3\mathbf{e}_3,$$

then by subtracting this equation from (4) and using **V**.1–9 we get

$$(a_1-b_1)\mathbf{e}_1+(a_2-b_2)\mathbf{e}_2+(a_3-b_3)\mathbf{e}_3=\mathbf{o}.$$

By the linear independence of $\mathbf{e}_1$, $\mathbf{e}_2$, $\mathbf{e}_3$, this can only be the trivial relation, i.e. $a_1=b_1$, $a_2=b_2$, $a_3=b_3$. This proves the following

Theorem. *If* $\mathbf{e}_1$, $\mathbf{e}_2$, $\mathbf{e}_3$ *are any three linearly independent vectors, then any vector* $\mathbf{a}$ *can be expressed as a linear combination of* $\mathbf{e}_1$, $\mathbf{e}_2$, $\mathbf{e}_3$ *with uniquely determined coefficients* a_1, a_2, a_3.

The vectors $\mathbf{e}_1$, $\mathbf{e}_2$, $\mathbf{e}_3$ are called a *set of basis-vectors* and a_1, a_2, a_3 are the *coordinates* of $\mathbf{a}$ relative to these basis-vectors. Occasionally we abbreviate equation (4) as

$$\mathbf{a}=\sum a_i\mathbf{e}_i,$$

where the summation sign $\sum$ indicates that the expression following it is summed over the values $i=1, 2, 3$.

If we are dealing with the same set of basis-vectors throughout, we may also abbreviate (4) as

$$\mathbf{a} \leftrightarrow (a_1, a_2, a_3).$$

If **b** is another vector, expressed as

$$\mathbf{b} \leftrightarrow (b_1, b_2, b_3)$$

in the same system, then we have

$$\mathbf{a}+\mathbf{b}=\sum a_i\mathbf{e}_i+\sum b_i\mathbf{e}_i$$
$$=\sum (a_i+b_i)\mathbf{e}_i,$$

hence

$$\mathbf{a}+\mathbf{b} \leftrightarrow (a_1+b_1,\ a_2+b_2,\ a_3+b_3); \tag{5}$$

similarly,

$$\lambda\mathbf{a} \leftrightarrow (\lambda a_1,\ \lambda a_2,\ \lambda a_3). \tag{6}$$

6. We now have all the tools at our disposal for setting up a coordinate-system in space.

We choose a point O as origin and take three linearly independent vectors $\mathbf{e}_1$, $\mathbf{e}_2$, $\mathbf{e}_3$. If P is any point in space, its position-vector relative to O may be expressed as a linear combination of the **e**'s:

$$OP=x_1\mathbf{e}_1+x_2\mathbf{e}_2+x_3\mathbf{e}_3.$$

The coordinates x_1, x_2, x_3 are called the *coordinates* of P relative to the basis-vectors $\mathbf{e}_1$, $\mathbf{e}_2$, $\mathbf{e}_3$ and the origin O. In this way every point of space is described by three real numbers. Conversely, any set of three real numbers (x_1, x_2, x_3) describes exactly one point in space: we merely form the linear combination $\mathbf{x}=x_1\mathbf{e}_1+x_2\mathbf{e}_2+x_3\mathbf{e}_3$ and apply the translation defined by **x** to the point O.

The lines through O in the direction of the basis-vectors are called the *coordinate-axes*; the axes defined by $\mathbf{e}_1$, $\mathbf{e}_2$, $\mathbf{e}_3$ are the 1-axis, 2-axis and 3-axis respectively. Any two axes define a *coordinate-plane*; thus the 12-plane is the plane containing the 1- and 2-axes, and the 13- and 23-planes are defined similarly.

7. In using coordinate-systems to solve problems one naturally adopts a system best suited to the particular problem, but there is one type of system which is of general importance: a rectangular coordinate-system. A coordin-

ate-system is said to be ***rectangular***, if the basis-vectors are mutually orthogonal vectors of unit-length. By contrast, a general coordinate-system is often described as ***oblique***.

In a rectangular system the end-points of the vectors $x_1\mathbf{e}_1$, $x_2\mathbf{e}_2$, $x_3\mathbf{e}_3$ (all starting at O) are the feet of the perpendiculars from the end-point of $\mathbf{x}$ to the coordinate-axes.

Thus $x_1\mathbf{e}_1$ is the projection of $\mathbf{x}$ on $\mathbf{e}_1$. Since $\mathbf{e}_1$ is a vector of unit-length, x_1 represents the ratio of this projection to 1, i.e. x_1 represents the projection of $\mathbf{x}$ on $\mathbf{e}_1$. We note in particular that x_1 is positive, negative or zero according as the angle between the directions of $\mathbf{x}$ and $\mathbf{e}_1$ is less than, greater than or equal to a right angle.

8. Let $\mathbf{a}$ be any vector; its length is a positive number or zero, which we denote by $|\mathbf{a}|$. Thus in writing $|\mathbf{a}|$ we ignore the direction of $\mathbf{a}$. The length of a vector has the following properties:

$$|\mathbf{a}|>0 \text{ unless } \mathbf{a}=\mathbf{o}, \tag{7}$$

and

$$|\lambda\mathbf{a}|=|\lambda|\,|\mathbf{a}|, \tag{8}$$

where $|\lambda|$ is the absolute value of λ (cf. footnote †, p. 4). Since the square of any real number is positive or zero, we have $|\lambda|^2=\lambda^2$, and so

$$|\lambda\mathbf{a}|^2=\lambda^2\,|\mathbf{a}|^2. \tag{9}$$

A vector of unit-length is often called a ***unit-vector***. In the next section we shall give an expression for the length of a vector in terms of its coordinates in a general coordinate-system. For the moment we shall confine ourselves to giving this expression in the special case of a rectangular coordinate-system.

If $\mathbf{e}_1$, $\mathbf{e}_2$, $\mathbf{e}_3$ are the basis-vectors of a rectangular coordinate-system and

$$\mathbf{a}=a_1\mathbf{e}_1+a_2\mathbf{e}_2+a_3\mathbf{e}_3,$$

then the segment OA representing $\mathbf{a}$ forms the diagonal of a rectangular parallelepiped† (cf. Fig. 4), whose edges lie

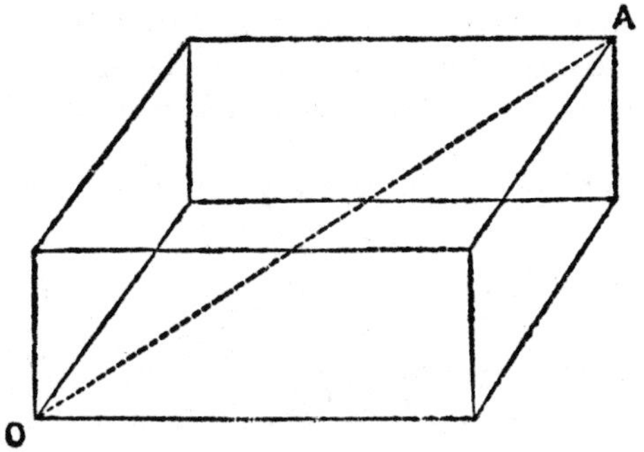

FIG. 4

along $\mathbf{e}_1$, $\mathbf{e}_2$, $\mathbf{e}_3$ and are of lengths a_1, a_2, a_3. The square of the length of this diagonal is found by a double application of Pythagoras' theorem, to be the sum of the squares of the edge-lengths, i.e.

$$|\mathbf{a}|^2 = a_1^2 + a_2^2 + a_3^2. \tag{10}$$

9. With any two vectors **a** and **b** we associate a scalar **a** . **b** which is defined by the equation

$$\mathbf{a} \,.\, \mathbf{b} = \tfrac{1}{2}(|\mathbf{a}+\mathbf{b}|^2 - |\mathbf{a}|^2 - |\mathbf{b}|^2). \tag{11}$$

This expression **a** . **b** is called the *scalar product* of **a** and **b**. It is linear in each factor ('bilinear') and commutative, i.e.

$$(\lambda\mathbf{a}+\mu\mathbf{b}) \,.\, \mathbf{c} = \lambda\mathbf{a} \,.\, \mathbf{c} + \mu\mathbf{b} \,.\, \mathbf{c}, \tag{12}$$

$$\mathbf{a} \,.\, (\lambda\mathbf{b}+\mu\mathbf{c}) = \lambda\mathbf{a} \,.\, \mathbf{b} + \mu\mathbf{a} \,.\, \mathbf{c}, \tag{13}$$

$$\mathbf{a} \,.\, \mathbf{b} = \mathbf{b} \,.\, \mathbf{a}. \tag{14}$$

Moreover, the length of a vector may be expressed in terms of the scalar product by the equation

$$|\mathbf{a}| = \sqrt{\mathbf{a} \,.\, \mathbf{a}}. \tag{15}$$

† A *parallelepiped* is a solid bounded by three pairs of parallel faces (analogously to a parallelogram in the plane). When adjacent faces cut at right angles the parallelepiped is called *rectangular*; a typical example is a brick.

The rules (14) and (15) are almost immediate consequences of the definition (11) (and (7) and (9)). To prove (12) and (13) we first express **a** . **b** in terms of coordinates referred to a rectangular coordinate-system. In such a system, if $\mathbf{a}=\sum a_i\mathbf{e}_i$ and $\mathbf{b}=\sum b_i\mathbf{e}_i$, then $\mathbf{a}+\mathbf{b}=\sum (a_i+b_i)\mathbf{e}_i$ and hence by (10) and (11),

$$\mathbf{a}\,.\,\mathbf{b}=\tfrac{1}{2}(\sum (a_i+b_i)^2-\sum a_i^2-\sum b_i^2), \text{ i.e.}$$

$$\mathbf{a}\,.\,\mathbf{b}=a_1b_1+a_2b_2+a_3b_3. \qquad (16)$$

This equation defines the scalar product in terms of the coordinates in a rectangular coordinate-system and from it we can now verify (12) and (13) without any difficulty, simply by writing out both sides. This may be left to the reader.

Equation (16) may also be used to give a geometrical interpretation of the scalar product. Firstly we note that (16) holds in any rectangular coordinate-system, because this is true of (10), from which (16) was derived. Now given any two vectors **a** and **b**, we can choose a rectangular coordinate-system in which the first basis-vector $\mathbf{e}_1$ lies in the same direction as **a**. Then $\mathbf{a}=|\mathbf{a}|\,.\,\mathbf{e}_1$ and so $\mathbf{a}\,.\,\mathbf{b}=|\mathbf{a}|\,.\,b_1$; here b_1 represents the projection of **b** on the direction of **a**, that is, $b_1=|\mathbf{b}|\cos\theta$, where θ is the angle between **a** and **b** (Fig. 5). Thus

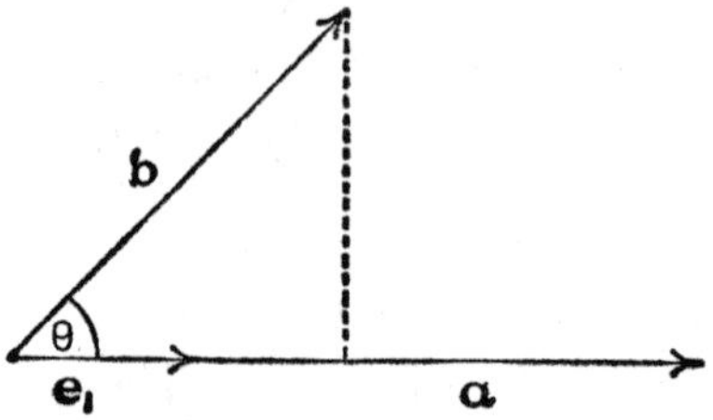

FIG. 5

$$\mathbf{a}\,.\,\mathbf{b}=|\mathbf{a}|\,|\mathbf{b}|\cos\theta. \qquad (17)$$

We note that **a** . **b** is positive when the directions of **a** and **b**

make an acute angle and negative when they make an obtuse angle. In particular: $\mathbf{a}\,.\,\mathbf{b}=0$ if $\mathbf{a}=\mathbf{o}$ or $\mathbf{b}=\mathbf{o}$ or if $\mathbf{a}$ and $\mathbf{b}$ are at right angles to each other. Thus it is possible for $\mathbf{a}\,.\,\mathbf{b}$ to be zero even though neither $\mathbf{a}$ nor $\mathbf{b}$ is the zero-vector. Another important difference from ordinary products is that in the product $\mathbf{a}\,.\,\mathbf{b}$ the factors are *vectors*, but the result is a *scalar*.

We can now give the expression for the length of a vector in a general (oblique) coordinate-system (corresponding to (10) for rectangular systems). Let $\mathbf{e}_1$, $\mathbf{e}_2$, $\mathbf{e}_3$ be the basis-vectors and write

$$\gamma_{ij}=\mathbf{e}_i\,.\,\mathbf{e}_j \quad (i, j=1, 2, 3). \tag{18}$$

Of the nine constants γ_{ij}, only six are independent, because $\gamma_{ij}=\gamma_{ji}$; the γ's are determined by the lengths of the vectors $\mathbf{e}_1$, $\mathbf{e}_2$, $\mathbf{e}_3$ and the angles between any two of these vectors (cf. (17)). Using (12) and (13) we now find that for vectors $\mathbf{a}$, $\mathbf{b}$ with coordinates given by $\mathbf{a}=\sum a_i\mathbf{e}_i$, $\mathbf{b}=\sum b_i\mathbf{e}_i$, we have

$$\mathbf{a}\,.\,\mathbf{b}=\sum \gamma_{ij}a_ib_j, \tag{19}$$

and in particular

$$|\mathbf{a}|^2=\mathbf{a}\,.\,\mathbf{a}=\sum \gamma_{ij}a_ia_j.$$

Written out in full this equation reads

$$|\mathbf{a}|^2=\gamma_{11}a_1^2+\gamma_{22}a_2^2+\gamma_{33}a_3^2+2\gamma_{12}a_1a_2+2\gamma_{13}a_1a_3+2\gamma_{23}a_2a_3. \tag{20}$$

This is the generalization of (10) to general coordinate-systems. In a rectangular system we have

$$\mathbf{e}_i\,.\,\mathbf{e}_j=\delta_{ij},$$

where†

$$\delta_{ij}=\begin{Bmatrix}1 \text{ if } i=j\\ 0 \text{ if } i\neq j\end{Bmatrix}. \tag{21}$$

With these values for γ_{ij}, (20) reduces to the equation (10). Because of the simple forms which (19) and (20) take in rectangular coordinate-systems (cf. (16) and (10)) we shall usually choose our system to be rectangular in any

† This symbol is known as the *Kronecker delta.*

problem involving scalar products or the lengths of vectors.

10. A second type of product which is often used is the *vector product* of two vectors. This arises when we determine the common perpendicular to two vectors. Let the vectors be **a** and **b**, then the vector **x** is at right angles to both **a** and **b** if $\mathbf{a} \cdot \mathbf{x} = \mathbf{b} \cdot \mathbf{x} = 0$. If in a rectangular coordinate-system, $\mathbf{a} \leftrightarrow (a_1, a_2, a_3)$ and $\mathbf{b} \leftrightarrow (b_1, b_2, b_3)$, then the co-ordinates (x_1, x_2, x_3) of **x** have to satisfy the equations

$$\left.\begin{aligned} a_1x_1 + a_2x_2 + a_3x_3 = 0, \\ b_1x_1 + b_2x_2 + b_3x_3 = 0. \end{aligned}\right\} \tag{22}$$

Solving these equations for x_1, x_2, x_3, we obtain

$$\frac{x_1}{a_2b_3 - a_3b_2} = \frac{x_2}{a_3b_1 - a_1b_3} = \frac{x_3}{a_1b_2 - a_2b_1}, \tag{23}$$

i.e.,

$$\begin{aligned} x_1 &= \lambda(a_2b_3 - a_3b_2), \\ x_2 &= \lambda(a_3b_1 - a_1b_3), \\ x_3 &= \lambda(a_1b_2 - a_2b_1), \end{aligned} \tag{24}$$

where λ is an arbitrary scalar. This is the complete solution of the system (22) provided that this system is of rank two, i.e. provided that **a** and **b** are linearly independent.† It will be recalled (LE, p. 68), that this is the case when at least one of the denominators in (23) is different from zero.

Let us denote by $\mathbf{a} \wedge \mathbf{b}$ the vector whose coordinates are the terms multiplying λ in (24), i.e.

$$\mathbf{a} \wedge \mathbf{b} \leftrightarrow (a_2b_3 - a_3b_2,\ a_3b_1 - a_1b_3,\ a_1b_2 - a_2b_1). \tag{25}$$

Then we can state our result as follows: For any two vectors **a** and **b** in three dimensions we can form the vector $\mathbf{a} \wedge \mathbf{b}$ defined by (25), such that $\mathbf{a} \wedge \mathbf{b} = \mathbf{o}$ if and only if **a** and **b** are linearly dependent, and in case **a** and **b** are independent, every vector orthogonal to both **a** and **b** is of the form $\lambda(\mathbf{a} \wedge \mathbf{b})$.

The result also has an immediate geometrical interpreta-

† See the present author's *Linear Equations* (p. 42) in this series, hereafter referred to as LE.

tion. To say that **a** and **b** are linearly independent just amounts to saying that **a** and **b** are not parallel, and so **a** and **b** define a plane in space; now the above result shows that there is just one direction in space perpendicular to this plane.

The vector $\mathbf{a} \wedge \mathbf{b}$ is called the *vector product* of **a** and **b**. Its main properties are again bilinearity,

$$\left.\begin{aligned}(\alpha\mathbf{a}+\beta\mathbf{b}) \wedge \mathbf{c} &= \alpha\mathbf{a} \wedge \mathbf{c}+\beta\mathbf{b} \wedge \mathbf{c},\\ \mathbf{a} \wedge (\alpha\mathbf{b}+\beta\mathbf{c}) &= \alpha\mathbf{a} \wedge \mathbf{b}+\beta\mathbf{a} \wedge \mathbf{c},\end{aligned}\right\} \tag{26}$$

and *anticommutativity*, expressed by the equation

$$\mathbf{a} \wedge \mathbf{b} = -\mathbf{b} \wedge \mathbf{a}; \tag{27}$$

in particular we also have the relation

$$\mathbf{a} \wedge \mathbf{a} = \mathbf{o}. \tag{28}$$

These relations are an immediate consequence of the definition (25), and their verification may be left to the reader. To obtain the length of the vector $\mathbf{a} \wedge \mathbf{b}$, we have, in a rectangular coordinate-system,

$$\begin{aligned}|\mathbf{a} \wedge \mathbf{b}|^2 &= (a_2b_3-a_3b_2)^2+(a_3b_1-a_1b_3)^2+(a_1b_2-a_2b_1)^2\\ &= a_2^2b_3^2+a_3^2b_2^2+a_3^2b_1^2+a_1^2b_3^2+a_1^2b_2^2+a_2^2b_1^2\\ &\quad -2(a_2b_2a_3b_3+a_1b_1a_3b_3+a_1b_1a_2b_2)\\ &= (a_1^2+a_2^2+a_3^2)(b_1^2+b_2^2+b_3^2)\\ &\quad -(a_1^2b_1^2+a_2^2b_2^2+a_3^2b_3^2+2a_1b_1a_2b_2+2a_1b_1a_3b_3\\ &\quad +2a_2b_2a_3b_3)\\ &= |\mathbf{a}|^2\,|\mathbf{b}|^2-(\mathbf{a}\,.\,\mathbf{b})^2.\end{aligned}$$

Hence, by (17),

$$\begin{aligned}|\mathbf{a} \wedge \mathbf{b}|^2 &= |\mathbf{a}|^2\,|\mathbf{b}|^2-|\mathbf{a}|^2\,|\mathbf{b}|^2\cos^2\theta\\ &= |\mathbf{a}|^2\,|\mathbf{b}|^2\sin^2\theta,\end{aligned}$$

i.e.,

$$|\mathbf{a} \wedge \mathbf{b}| = |\mathbf{a}|\,|\mathbf{b}|\,|\sin\theta|, \tag{29}$$

where θ is the angle between **a** and **b**. Thus the length of the vector $\mathbf{a} \wedge \mathbf{b}$ is the area of the parallelogram with sides given by **a** and **b**.

Although the vector product was defined in terms of a particular coordinate-system, we see from (29) that its length is independent of the choice of coordinate-system (as long as it is a rectangular one). We already know that $\mathbf{a} \wedge \mathbf{b}$ is at right angles to $\mathbf{a}$ and $\mathbf{b}$; therefore if (25) is applied in different rectangular coordinate-systems, the resulting vectors can differ at most by a factor -1. In Chapter Three we shall restrict the coordinate-systems (to right-handed rectangular systems) so as to avoid this ambiguity of sign. At the same time we shall show how to compute the vector product in general coordinate-systems.

EXERCISES ON CHAPTER ONE

1. For any vectors $\mathbf{a}$ and $\mathbf{b}$ show that
$$(\mathbf{a}+\mathbf{b})^2+(\mathbf{a}-\mathbf{b})^2=2(\mathbf{a}^2+\mathbf{b}^2).$$

2. In a triangle with sides a, b, c and corresponding angles α, β, γ show that
$$a^2+b^2-c^2=2ab\cos\gamma.$$
[Hint: use (11) and (17).]

3. Show that the points (5, −5, 4), (3, −4, 0), (5, 4, 1) (in a rectangular system) form a right-angled triangle in space and find the fourth point to complete the rectangle.

4. Find the circumcentre of the triangle with the vertices O, (3, 1, 2), (1, 2, −3) (in a rectangular system). Find also the radius of the circumcircle.
[Express the fact that the circumcentre lies in the plane of the triangle on the perpendicular bisectors of the sides.]

5. Show that in a tetrahedron the joins of mid-points of opposite sides meet in a point which bisects each of them.
[Let the vertices be A, B, C, D with position-vectors $\mathbf{a}$, $\mathbf{b}$, $\mathbf{c}$, $\mathbf{d}$ respectively. Then the mid-points of AB and CD are $\frac{1}{2}(\mathbf{a}+\mathbf{b})$ and $\frac{1}{2}(\mathbf{c}+\mathbf{d})$ respectively, and the mid-point of their join is $\frac{1}{4}(\mathbf{a}+\mathbf{b}+\mathbf{c}+\mathbf{d})$.]

6. If $\mathbf{a}_1$, $\mathbf{a}_2$, $\mathbf{a}_3$ are three linearly independent vectors, show that there is just one set of vectors $\mathbf{b}_1$, $\mathbf{b}_2$, $\mathbf{b}_3$ such that $\mathbf{a}_i \,.\, \mathbf{b}_j=\delta_{ij}$ (cf. (21)), and show that $\mathbf{x}=\Sigma(\mathbf{x} \,.\, \mathbf{b}_i)\mathbf{a}_i$, for any vector $\mathbf{x}$.

7. If $\mathbf{a}_1$, $\mathbf{a}_2$, $\mathbf{a}_3$ are three linearly independent vectors, find a set of mutually orthogonal unit-vectors $\mathbf{e}_1$, $\mathbf{e}_2$, $\mathbf{e}_3$ such that $\mathbf{e}_1$ is linearly dependent on $\mathbf{a}_1$ and $\mathbf{e}_2$ on $\mathbf{a}_1$ and $\mathbf{a}_2$.
[Note that for any orthogonal unit-vectors $\mathbf{e}_1$, $\mathbf{e}_2$ and any vector $\mathbf{x}$, $\mathbf{x}-(\mathbf{x} \,.\, \mathbf{e}_1)\mathbf{e}_1$ is orthogonal to $\mathbf{e}_1$ and $\mathbf{x}-(\mathbf{x} \,.\, \mathbf{e}_1)\mathbf{e}_1-(\mathbf{x} \,.\, \mathbf{e}_2)\mathbf{e}_2$ is orthogonal to $\mathbf{e}_1$ and $\mathbf{e}_2$.]

CHAPTER TWO

Lines and Planes

1. The different lines through a given point, O say, of space differ only in their directions and may therefore be specified by vectors along them. A given line l through O is completely defined by laying a non-zero vector **a** along it; the vector **a** determines l, but **a** is determined only up to a non-zero scalar multiple by l. This means that l determines and is determined by the ratios $a_1 : a_2 : a_3$ of the coordinates of **a** (in some coordinate-system). For this reason the numbers a_1, a_2, a_3 are called the *direction-ratios* of the line l (in the given coordinate-system).

Now strictly speaking there are two directions defined on each line l; if one of them is given by the vector **a**, the other is given by $-\mathbf{a}$, and the vector $\lambda\mathbf{a}$ defines one or the other direction according as λ is positive or negative. For definiteness we shall single out one of these two directions as the *positive direction* and denote the line with this direction by $\vec{l}$; we say that l has been *oriented*. In this way every line gives rise to two *oriented lines*† which consist of of the same line but with opposite directions. Thus an oriented line may be thought of as a line with an arrow along it (pointing in the positive direction, say). The two orientations on l may also be defined by the two unit-vectors along l. If **a** is a non-zero vector along $\vec{l}$, then the unit-vector **u** describing $\vec{l}$ is given by

$$\mathbf{u} = \frac{1}{|\mathbf{a}|}\,\mathbf{a}. \qquad (1)$$

† This name seems preferable to the older term 'directed line' and more in accordance with current terminology.

Let **u** be a unit-vector corresponding to the oriented line $\vec{l}$ through O, and take a rectangular coordinate-system with origin at O. By taking the scalar product of **u** with the basis-vectors $\mathbf{e}_1$, $\mathbf{e}_2$ and $\mathbf{e}_3$ in turn, we find

$$u_i = \mathbf{u} \,.\, \mathbf{e}_i = \cos \alpha_i \quad (i=1, 2, 3) \tag{2}$$

where α_i is the angle which $\vec{l}$ makes with the direction of $\mathbf{e}_i$. The angles α_1, α_2, α_3 are called the *direction-angles* and their cosines the *direction-cosines* of $\vec{l}$. Thus an oriented line through O may be determined by its direction-angles, or by its direction-cosines, or, ignoring orientation, by its direction-ratios.

Any three numbers, not all zero, represent the direction-ratios of some line, since every non-zero vector determines a line. If **a** is a non-zero vector, and $\vec{l}$ the oriented line through O determined by **a**, we obtain the direction-cosines of $\vec{l}$ by *normalizing* **a**, i.e. dividing it by the scalar $|\mathbf{a}|$ (as in (1)), in order to obtain a unit-vector in the direction of $\vec{l}$. Explicitly the coordinates of this unit-vector **u** (in a rectangular coordinate-system) are

$$u_i = a_i(a_1^2 + a_2^2 + a_3^2)^{-1/2} \quad (i=1, 2, 3).$$

E.g. the line with the direction-ratios $1:2:-3$ has direction-cosines $1/\sqrt{14}$, $2/\sqrt{14}$, $-3/\sqrt{14}$ and angles 74° . 12 56° . 62 and 145° . 58.† The same line differently oriented' has direction-cosines $-1/\sqrt{14}$, $-2/\sqrt{14}$, $3/\sqrt{14}$ and direction-angles 105° . 88, 123° . 38 and 34° . 42.

Any three numbers u_1, u_2, u_3 such that $u_1^2 + u_2^2 + u_3^2 = 1$, may be taken as the direction-cosines of some line, namely the (oriented) line in the direction of the vector $\mathbf{u} = \sum u_i \mathbf{e}_i$. From the interpretation of the u_i as cosines we see that the angles α_1, α_2, α_3 which a line makes with the basis-vectors (of a rectangular system) satisfy the relation

† Note that a direction-angle is uniquely determined by its cosine because it must lie between 0° and 180°, being less than or greater than 90° according as its cosine is positive or negative.

$$\cos^2 \alpha_1 + \cos^2 \alpha_2 + \cos^2 \alpha_3 = 1. \tag{3}$$

Conversely, any three angles whose cosines satisfy this relation, form the direction-angles of an oriented line.

2. Next we wish to describe the different points on a line l. We suppose again that l passes through the origin O of the coordinate-system (which need not be rectangular). If $\mathbf{a}$ is any non-zero vector along l, then the points represented by the vectors $\lambda\mathbf{a}$ (where λ is a scalar) lie on l and we obtain all the points of the line l by varying λ. We also say: l is *spanned* by the vector $\mathbf{a}$. Thus the equation

$$\mathbf{x} = \lambda\mathbf{a} \tag{4}$$

may be regarded as an equation for the line l. Here $\mathbf{a}$ is any non-zero vector along l; the scalar λ is called the *parameter* of the line and (4) is *the parametric form* of the equation of the line.

In order to represent a line l which does not pass through O we choose any point P on l; let $\mathbf{p}$ be the vector OP and let $\mathbf{a}$ be a non-zero vector in the direction of l. Then for any scalar λ, $\mathbf{p} + \lambda\mathbf{a}$ represents a point on l, and any point on l may be represented in this way (Fig. 6). Thus an equation for the line is

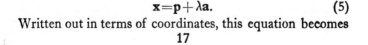

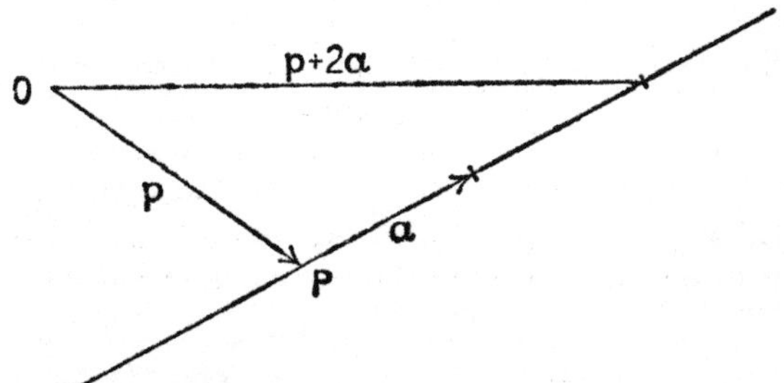

FIG. 6

$$\mathbf{x} = \mathbf{p} + \lambda\mathbf{a}. \tag{5}$$

Written out in terms of coordinates, this equation becomes

$$\left.\begin{aligned} x_1 &= p_1 + \lambda a_1, \\ x_2 &= p_2 + \lambda a_2, \\ x_3 &= p_3 + \lambda a_3, \end{aligned}\right\} \tag{6}$$

which on elimination of the parameter λ leads to the form

$$\frac{x_1 - p_1}{a_1} = \frac{x_2 - p_2}{a_2} = \frac{x_3 - p_3}{a_3} \; (= \lambda) \tag{7}$$

for the scalar equations of the straight line through the point P with coordinates (p_1, p_2, p_3) and with direction-ratios (a_1, a_2, a_3).

For any particular line, $p_1, p_2, p_3, a_1, a_2, a_3$ are given constants, while x_1, x_2, x_3 are the coordinates of the general point on the line. It is then convenient to abbreviate the equations (6) by

$$\mathbf{x} \leftrightarrow (p_1, p_2, p_3) + \lambda(a_1, a_2, a_3), \tag{8}$$

(in the notation of 1.5) where of course a fixed coordinate-system is understood. For example, the line through the point $(-2, 0, 5)$ with the direction-ratios $1 : 2 : -3$ is given by

$$\mathbf{x} \leftrightarrow (-2, 0, 5) + \lambda(1, 2, -3),$$

or in scalar form,

$$\frac{x_1 + 2}{1} = \frac{x_2}{2} = \frac{x_3 - 5}{-3}.$$

A line may also be specified by two of its points and it is not difficult to obtain an equation for it from these data. Thus let P and Q be two distinct points on the line l, whose position-vectors (relative to O) are $\mathbf{p}$ and $\mathbf{q}$ respectively. Since P and Q are distinct, $\mathbf{q} - \mathbf{p} \neq \mathbf{o}$, and $\mathbf{q} - \mathbf{p}$ is a vector in the direction of l. Hence an equation for l is

$$\mathbf{x} = \mathbf{p} + \lambda(\mathbf{q} - \mathbf{p}). \tag{9}$$

As an example, the line through the points $(4, -1, 2)$ and $(3, 1, 0)$ is given by

$$\mathbf{x} \leftrightarrow (4, -1, 2) + \lambda(-1, 2, -2).$$

3. To represent the points in a plane through O we take

two linearly independent vectors in the plane, **a** and **b** say, and form all linear combinations of **a** and **b**. In this way we obtain as an equation for the plane

$$\mathbf{x}=\lambda\mathbf{a}+\mu\mathbf{b}, \tag{10}$$

where λ and μ are scalar parameters, the 'coordinates' of the plane relative to **a** and **b** as basis-vectors. To represent a general plane we take any two linearly independent vectors **a**, **b** defining directions in the plane and the vector **p** of any point P of the plane, relative to O. The general point of the plane is then

$$\mathbf{x}=\mathbf{p}+\lambda\mathbf{a}+\mu\mathbf{b}. \tag{11}$$

Conversely, any equation of the form (11), where **a** and **b** are linearly independent, represents a plane. We shall refer to it as the plane through P *spanned* by the vectors **a** and **b**.

If P, Q, R are three non-collinear points with vectors **p**, **q**, **r** respectively, then $\mathbf{q}-\mathbf{p}$ and $\mathbf{r}-\mathbf{p}$ are linearly independent vectors; therefore an equation for the plane through P, Q and R is

$$\mathbf{x}=\mathbf{p}+\lambda(\mathbf{q}-\mathbf{p})+\mu(\mathbf{r}-\mathbf{p}).$$

For this is clearly satisfied by **p**, **q** and **r**.

The equation (11) of the plane may be brought to a more convenient form by eliminating the parameters λ and μ. Let **u** be a unit-vector perpendicular to the plane. Taking the scalar product of (11) with **u** we find

$$\mathbf{u}\,.\,\mathbf{x}=\mathbf{u}\,.\,\mathbf{p}, \tag{12}$$

the other terms vanishing because **a** and **b** lie in the plane and so are perpendicular to **u**. It follows that every vector **x** satisfying (11) must satisfy (12). Conversely, if **x** satisfies (12), then $\mathbf{x}-\mathbf{p}$ is perpendicular to **u** and hence lies in the plane, therefore it is a linear combination of **a** and **b**. Thus we have shown that (11) is equivalent to (12), so that (12) may also be regarded as an equation for the plane. To obtain a unit-vector perpendicular to the plane we may form $\mathbf{a}\wedge\mathbf{b}$ and divide by $|\mathbf{a}\wedge\mathbf{b}|$, which does not vanish because **a** and **b** are linearly independent.

Assume now that we have a rectangular coordinate-system. Using coordinates, we can write (12) as

$$u_1x_1+u_2x_2+u_3x_3=k, \tag{13}$$

where $k=\mathbf{u}\,.\,\mathbf{p}$. The coefficients in (13) may be interpreted as follows (Fig. 7): since $\mathbf{u}$ is a unit-vector perpendicular

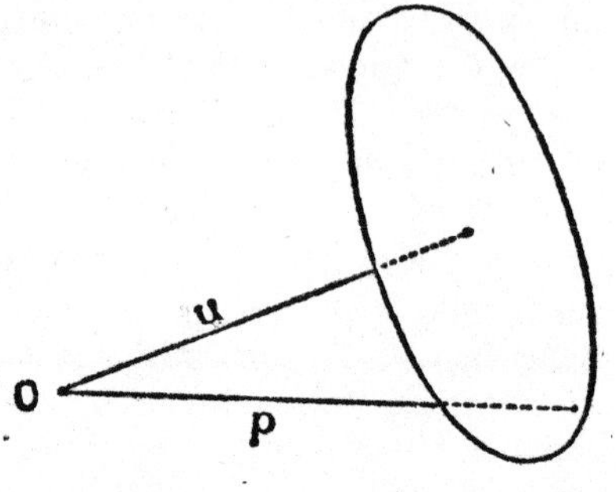

FIG. 7

to the plane, its coordinates u_1, u_2, u_3 are the direction-cosines of the perpendicular. Further, $\mathbf{u}\,.\,\mathbf{p}$ represents the projection of $\mathbf{p}$ on $\mathbf{u}$, i.e. the distance of the origin from the plane. The unique direction perpendicular to the plane is also called the direction *normal to the plane.*

The only restriction on the coefficients in (13) is that $u_1^2+u_2^2+u_3^2=1$. This shows that any equation

$$c_1x_1+c_2x_2+c_3x_3=c, \tag{14}$$

where c_1, c_2, c_3 are not all zero, represents a plane. For on dividing the whole equation by $\sqrt{c_1^2+c_2^2+c_3^2}$ we can reduce it to the form (13). As an example, the equation

$$2x_1+x_2-5x_3=3$$

may be brought to the form (13) by dividing by $\sqrt{2^2+1^2+5^2}=\sqrt{30}$. Hence it represents the plane whose normal has direction-cosines $2/\sqrt{30}$, $1/\sqrt{30}$, $-5/\sqrt{30}$ and which is at a distance $3/\sqrt{30}=\sqrt{3/10}$ from the origin.

If in equation (14), $c \neq 0$, we may divide by c and so bring it to the form

$$a_1x_1+a_2x_2+a_3x_3=1. \tag{15}$$

This is called the *intercept form* of the equation, because the coefficients are the reciprocals of the intercepts of the plane with the coordinate-axes. Thus the intersection of the plane with the 1-axis is given by putting $x_2=x_3=0$ in (15), i.e.

$$a_1x_1=1. \tag{16}$$

If $a_1 \neq 0$, this has the solution $x_1=1/a_1$, thus the plane meets the 1-axis in the point $(1/a_1, 0, 0)$ in case $a_1 \neq 0$. If $a_1=0$, (16) has no solution, i.e. the plane does not meet the 1-axis. In this case the plane is parallel to the 1-axis. The same argument shows that the plane meets the 2- and 3-axes, if at all, in the points $(0, 1/a_2, 0)$ and $(0, 0, 1/a_3)$ respectively. Of course if in (14) $c=0$, the equation cannot be brought to the form (15). In this case the plane passes through O, so that its intercept with each axis is O, and therefore not enough to determine the plane.

4. Two lines in space which intersect, lie in a plane and one can therefore speak of the angle between the lines. However, in general two lines in space need not intersect. They may be parallel, in which case they still lie in a plane; or they may neither intersect nor be parallel. In the latter case they are called *skew*. E.g., any one of the twelve edges of a cube meets four other edges and is parallel to three others and skew to the remaining four.

If we have two general lines l, m in space, we define the angle between them as the angle between the lines l', m' which pass through O (and hence intersect) and are parallel to l and m respectively. This angle is uniquely determined if we restrict it to lie between 0° and 90°. If on the other hand we are dealing with oriented lines the angle lies between 0° and 180° and is unique in this range. Let the equations of l and m be

$$\mathbf{x}=\mathbf{p}+\lambda\mathbf{u},$$
$$\mathbf{x}=\mathbf{q}+\mu\mathbf{v},$$

where **u** and **v** are unit-vectors.

If we take the positive directions along l and m to be defined by **u** and **v** respectively, we obtain two oriented lines, $\overrightarrow{l}$ and $\overrightarrow{m}$ say, whose direction-cosines (in a rectangular coordinate-system) are given by the coordinates of **u** and **v** respectively. The angle between l and m is by definition the angle between **u** and **v**. Hence if this angle is denoted by θ, then

$$\cos\theta=\mathbf{u}\,.\,\mathbf{v}.$$

Of course the angle, θ_0 say, between the unoriented lines l and m is θ or $180°-\theta$, whichever does not exceed $90°$. Explicitly we have

$$\cos\theta=u_1v_1+u_2v_2+u_3v_3,$$

or, if α_1, α_2, α_3 and β_1, β_2, β_3 are the direction-angles of the two lines,

$$\cos\theta=\cos\alpha_1\cos\beta_1+\cos\alpha_2\cos\beta_2+\cos\alpha_3\cos\beta_3.$$

In particular, if $\mathbf{u}\,.\,\mathbf{v}=0$, the lines are at right angles, though they do not necessarily meet, and if $\mathbf{u}\,.\,\mathbf{v}=\pm1$, the lines are parallel (and hence lie in the same plane).

For example, consider the two lines

$$\mathbf{x}\leftrightarrow(2, 3, 5)+\lambda(1, 2, -1) \text{ and}$$
$$\mathbf{x}\leftrightarrow(4, -8, 0)+\mu(3, -1, 4).$$

The cosine of the angle between them is

$$\frac{-3}{\sqrt{6}\,\sqrt{26}} \doteq -0.240,$$

whence the angle between the (unoriented) lines is $76°\ 6'$.

5. Suppose we are given a line and a plane in space. If we visualize this situation we see that either

(i) the line and the plane meet in a single point, or

(ii.a) the line lies parallel to, but not in the plane,

(ii.b) the line lies entirely in the plane.

We shall now derive this result algebraically, using the

equations of the line and plane obtained in Nos. 2 and 3. The algebraic method has the advantage that it is applicable even if the number of dimensions exceeds three; on the other hand the method is to some extent illuminated by the geometrical interpretation given above, even when the equations arise in a different context (cf. LE, p. 48 f.).

We take the equation of the line in the form

$$\mathbf{x}=\mathbf{p}+\lambda\mathbf{a} \quad (\mathbf{a}\neq\mathbf{o}), \tag{17}$$

and the plane

$$\mathbf{x}\,.\,\mathbf{c}=k \quad (\mathbf{c}\neq\mathbf{o}). \tag{18}$$

The intersection is given by those vectors $\mathbf{x}$ which satisfy both (17) and (18). Now the vector $\mathbf{x}$ given by (17) satisfies (18) if and only if

$$(\mathbf{p}+\lambda\mathbf{a})\,.\,\mathbf{c}=k, \text{ i.e.}$$

$$\mathbf{p}\,.\,\mathbf{c}+\lambda\mathbf{a}\,.\,\mathbf{c}=k. \tag{19}$$

This is a linear equation for λ and its solution, if it exists, gives the intersection.

(i) $\mathbf{a}\,.\,\mathbf{c}\neq 0$. This means that the direction of the line (namely $\mathbf{a}$) is not perpendicular to the normal (direction: $\mathbf{c}$) of the plane, in other words, the line is not parallel to the plane. In this case (19) has a unique solution

$$\lambda=(k-\mathbf{p}\,.\,\mathbf{c})/(\mathbf{a}\,.\,\mathbf{c}).$$

Inserting this value for λ in (17) we obtain the unique point of intersection.

(ii) $\mathbf{a}\,.\,\mathbf{c}=0$. Now the line is parallel to the plane.

Either (a) $\mathbf{p}\,.\,\mathbf{c}\neq k$, then no value of λ satisfies (19), and so the line and plane have no common point, or (b) $\mathbf{p}\,.\,\mathbf{c}=k$; then every value of λ satisfies (19), i.e. the line lies entirely in the plane.

We have thus obtained the intersection by algebraic means. At the same time we see how in a simple situation the typical distinction of cases arises (LE, p. 1).

6. If we are given two planes in space, we expect them (on geometrical grounds) either (i) to intersect in a line, or (ii.a) to be parallel but not coincident, or (ii.b) to coincide.

Let us verify this algebraically. We may take the equations for the planes in the form

$$\mathbf{x}\,.\,\mathbf{a}=k,\quad \mathbf{x}\,.\,\mathbf{b}=l,\quad (\mathbf{a}\neq\mathbf{o},\ \mathbf{b}\neq\mathbf{o}). \tag{20}$$

In case (i) the planes are not parallel, i.e. their normals are not parallel, so that **a** and **b** are linearly independent. We show that there is then a point common to the planes whose position-vector is of the special form $\alpha\mathbf{a}+\beta\mathbf{b}$. This amounts to showing that the equations

$$\begin{aligned}(\xi\mathbf{a}+\eta\mathbf{b})\,.\,\mathbf{a}&=k\\(\xi\mathbf{a}+\eta\mathbf{b})\,.\,\mathbf{b}&=l\end{aligned} \tag{21}$$

can be solved for ξ, η. These equations may be written

$$\begin{aligned}\xi\mathbf{a}\,.\,\mathbf{a}+\eta\mathbf{b}\,.\,\mathbf{a}&=k\\\xi\mathbf{a}\,.\,\mathbf{b}+\eta\mathbf{b}\,.\,\mathbf{b}&=l;\end{aligned} \tag{22}$$

their determinant is

$$\begin{aligned}(\mathbf{a}\,.\,\mathbf{a})(\mathbf{b}\,.\,\mathbf{b})-(\mathbf{a}\,.\,\mathbf{b})^2&=|\mathbf{a}|^2|\mathbf{b}|^2(1-\cos^2\theta)\\&=|\mathbf{a}|^2|\mathbf{b}|^2\sin^2\theta,\end{aligned}$$

where θ is the angle between **a** and **b**. Since **a** and **b** are linearly independent, they are not zero and θ is neither 0° nor 180°; hence the determinant does not vanish and so the equations (22) have a unique solution, say

$$\mathbf{p}=\alpha\mathbf{a}+\beta\mathbf{b}. \tag{23}$$

Let **c** be any vector perpendicular to **a** and **b**. We shall show that the intersection of the planes consists of the line

$$\mathbf{x}=\mathbf{p}+\lambda\mathbf{c}, \tag{24}$$

where **p** is given by (23). By construction, $\mathbf{p}\,.\,\mathbf{a}=k$, $\mathbf{p}\,.\,\mathbf{b}=l$ and $\mathbf{c}\,.\,\mathbf{a}=\mathbf{c}\,.\,\mathbf{b}=0$, whence

$$\begin{aligned}(\mathbf{p}+\lambda\mathbf{c})\,.\,\mathbf{a}&=\mathbf{p}\,.\,\mathbf{a}=k,\\(\mathbf{p}+\lambda\mathbf{c})\,.\,\mathbf{b}&=\mathbf{p}\,.\,\mathbf{b}=l,\end{aligned}$$

for all λ, which shows that every point of the line (24) belongs to the intersection of the two planes. Conversely, if $\mathbf{x}=\mathbf{q}$ is any point of the intersection then $\mathbf{q}\,.\,\mathbf{a}=k$, $\mathbf{q}\,.\,\mathbf{b}=l$, hence

$$(\mathbf{q}-\mathbf{p})\,.\,\mathbf{a}=(\mathbf{q}-\mathbf{p})\,.\,\mathbf{b}=0,$$

i.e. $\mathbf{q}-\mathbf{p}$ is perpendicular to both $\mathbf{a}$ and $\mathbf{b}$, and so $\mathbf{q}-\mathbf{p}$ is perpendicular to the plane spanned by $\mathbf{a}$ and $\mathbf{b}$. Now the definition of $\mathbf{c}$ shows that $\mathbf{q}-\mathbf{p}$ must be linearly dependent on $\mathbf{c}$, say $\mathbf{q}-\mathbf{p}=\lambda\mathbf{c}$. Thus $\mathbf{q}=\mathbf{p}+\lambda\mathbf{c}$, which shows that $\mathbf{q}$ lies on the line (24). This line is therefore the intersection of the given planes.

We note that from the algebraic point of view, $\mathbf{x}=\mathbf{p}$ is a particular solution of the system (20), and $\mathbf{x}=\mathbf{c}$ is a solution of the associated homogeneous system (cf. LE, p. 38). In terms of coordinates, (20) is a system of two equations in three unknowns (the coordinates of $\mathbf{x}$), and it is of rank two provided that $\mathbf{a}$ and $\mathbf{b}$ are linearly independent.

If the planes are parallel, $\mathbf{a}$ and $\mathbf{b}$ are linearly dependent, say $\mathbf{b}=\gamma\mathbf{a}$. The equations (20) then read

$$\mathbf{x}\,.\,\mathbf{a}=k,\quad \gamma\mathbf{x}\,.\,\mathbf{a}=l,$$

and it is clear that these equations either

(a) have no solution, namely if $l\neq\gamma k$, or

(b) reduce to the single equation $\mathbf{x}\,.\,\mathbf{a}=k$, if $l=\gamma k$.

Accordingly (a) the planes are parallel without a common point or (b) the planes coincide.

As an illustration of the general case we consider the planes

$$\mathbf{x}\,.\,\mathbf{a}=96,\quad \mathbf{x}\,.\,\mathbf{b}=18,$$

where $\mathbf{a}\leftrightarrow(-2, 7, -4)$ and $\mathbf{b}\leftrightarrow(1, 1, -1)$ (in rectangular coordinates). The equations (22) read

$$\begin{aligned}69\xi+9\eta&=96,\\ 9\xi+3\eta&=18.\end{aligned}\tag{25}$$

They have a unique solution, namely $\xi=1$, $\eta=3$, and hence

$$\mathbf{p}\leftrightarrow(-2, 7, -4)+3(1, 1, -1)=(1, 10, -7).$$

To obtain a vector $\mathbf{c}$ perpendicular to $\mathbf{a}$ and $\mathbf{b}$ we form $\mathbf{a}\wedge\mathbf{b}$; its coordinates are $(-3, -6, -9)$, and dividing by -3 (which simplifies the coordinates without affecting the direction) we get

$$\mathbf{c}\leftrightarrow(1, 2, 3).$$

Thus the direction-ratios of the line of intersection are 1 : 2 : 3, and the intersection is

$$\mathbf{x} \leftrightarrow (1, 10, -7) + \lambda(1, 2, 3).$$

7. As we saw in No. 4, two lines in space need not intersect nor even lie in the same plane. But we shall now show that there always exists a line which meets both the given lines and is perpendicular to both. Moreover, if the given lines are not parallel, there is exactly one such line.

Let l and m be the given lines, with equations

$$\mathbf{x} = \mathbf{p} + \lambda\mathbf{a} \quad (\mathbf{a} \neq \mathbf{o}), \tag{26}$$

$$\mathbf{x} = \mathbf{q} + \mu\mathbf{b} \quad (\mathbf{b} \neq \mathbf{o}). \tag{27}$$

The line-segment RS, where R is the point on l with position-vector given by (26) and S the point on m with position-vector given by (27) is represented by the vector

$$\mathbf{q} + \mu\mathbf{b} - \mathbf{p} - \lambda\mathbf{a},$$

and the condition that RS is perpendicular to both l and m is expressed by the equations

$$\begin{aligned} (\mathbf{q} - \mathbf{p} + \mu\mathbf{b} - \lambda\mathbf{a}) \,.\, \mathbf{a} &= 0, \\ (\mathbf{q} - \mathbf{p} + \mu\mathbf{b} - \lambda\mathbf{a}) \,.\, \mathbf{b} &= 0. \end{aligned} \tag{28}$$

These equations may be written

$$\begin{aligned} \lambda\mathbf{a} \,.\, \mathbf{a} - \mu\mathbf{b} \,.\, \mathbf{a} &= \mathbf{q} \,.\, \mathbf{a} - \mathbf{p} \,.\, \mathbf{a}, \\ \lambda\mathbf{a} \,.\, \mathbf{b} - \mu\mathbf{b} \,.\, \mathbf{b} &= \mathbf{q} \,.\, \mathbf{b} - \mathbf{p} \,.\, \mathbf{b}, \end{aligned} \tag{29}$$

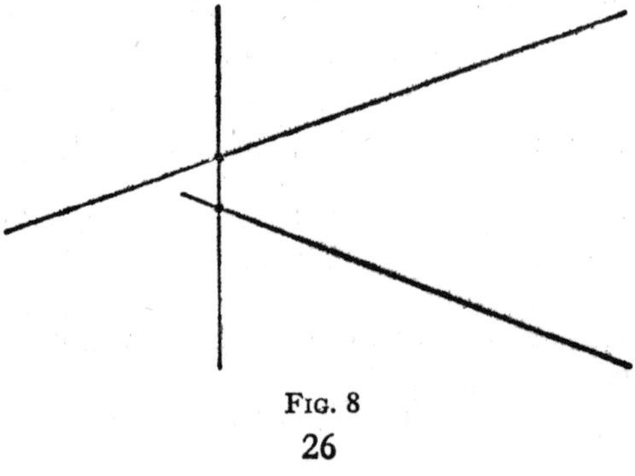

FIG. 8

and as we saw in No. 6 (cf. (23)), these equations have a unique solution if **a** and **b** are linearly independent, i.e. provided that l and m are not parallel. The solution gives the value of the parameters for the points in which the common perpendicular meets the two lines (Fig. 8). As an example, let l and m be given by

$$\mathbf{x} \leftrightarrow (-3, 3, 6) + \lambda(3, 0, -1),$$

and

$$\mathbf{x} \leftrightarrow (6, -4, 3) + \mu(4, -5, 2)$$

respectively (in rectangular coordinates). In this case

$$\mathbf{q} - \mathbf{p} + \mu\mathbf{b} - \lambda\mathbf{a} \leftrightarrow (9, -7, -3) + \mu(4, -5, 2) - \lambda(3, 0, -1),$$

and the equations (29) take the form

$$10\lambda - 10\mu = 30,$$
$$10\lambda - 45\mu = 65.$$

The solution is $\lambda = 2$, $\mu = -1$. Hence the points in which the common perpendicular meets l and m respectively are

$$(-3, 3, 6) + 2(3, 0, -1) = (3, 3, 4)$$

and

$$(6, -4, 3) + (-1)(4, -5, 2) = (2, 1, 1).$$

An equation for the common perpendicular is obtained by writing down an expression for the line through these two points:

$$\mathbf{x} \leftrightarrow (2, 1, 1) + \lambda(1, 2, 3).$$

When the lines are parallel they still have a common perpendicular, but it is no longer unique. The lines may then be taken to be

$$\mathbf{x} = \mathbf{p} + \lambda\mathbf{a} \text{ and } \mathbf{x} = \mathbf{q} + \mu\mathbf{a} \quad (\mathbf{a} \neq \mathbf{o});$$

then the equations (28) reduce to the single equation

$$(\mathbf{q} - \mathbf{p} + (\mu - \lambda)\mathbf{a}) \cdot \mathbf{a} = 0,$$

with the solution

$$\lambda - \mu = (\mathbf{q} - \mathbf{p}) \cdot \mathbf{a}/|\mathbf{a}|^2. \tag{30}$$

We may fix μ arbitrarily, say $\mu = \mu_0$, and λ is then uniquely determined by (30), which corresponds to the fact that there is one common perpendicular through each point of *m* (and of course one through each point of *l*).

E.g. if the lines are given by

$$\mathbf{x} \leftrightarrow (2, 1, 4) + \lambda(3, -1, 2)$$

and

$$\mathbf{x} \leftrightarrow (3, 4, 2) + \mu(3, -1, 2),$$

then

$$\lambda - \mu = -4/14 = -2/7.$$

Thus the perpendicular through the point $\mu = \mu_0$ of m meets m and l in the points $(3, 4, 2) + \mu_0(3, -1, 2)$ and $(2, 1, 4) + \left(\mu_0 - \frac{2}{7}\right)(3, -1, 2)$ respectively, and is given by $\mathbf{x} \leftrightarrow (3, 4, 2) + \mu_0(3, -1, 2) + \kappa\{(1, 3, -2) + 2/7(3, -1, 2)\}$, where κ is the parameter.

8. Let us again consider two lines in space which are not parallel. If the lines intersect, then their intersection is a single point and their common perpendicular clearly passes through this point. Thus we have a method of determining the point of intersection of two lines (when there is such an intersection). If the lines are skew, the common perpendicular represents the shortest distance between them. This may be seen as follows: Let the two lines be given by the equations (26) and (27) and let the common perpendicular be

$$\mathbf{x} = \mathbf{r} + \nu\mathbf{c} \quad (\mathbf{c} \neq \mathbf{o}).$$

Thus **c** is a vector perpendicular to both **a** and **b.** The lines l and m then lie in the planes

$$\mathbf{x} \cdot \mathbf{c} = \mathbf{p} \cdot \mathbf{c} \tag{31}$$

and

$$\mathbf{x} \cdot \mathbf{c} = \mathbf{q} \cdot \mathbf{c} \tag{32}$$

respectively, as follows from the fact that **a** and **b** are perpendicular to **c**. These planes (31) and (32) are parallel, therefore the distance from any point R in the plane (31) to any point S in the plane (32) is least precisely when RS is perpendicular to the two planes, i.e. when RS is parallel to **c**. But this is just the case when R and S are the feet of the common perpendicular to the planes.

EXERCISES ON CHAPTER TWO

Any coordinates written out in these exercises are taken to refer to a rectangular coordinate-system.

1. A cube is placed with its sides parallel to the axes, one vertex at the origin and the opposite vertex at $P \leftrightarrow (1, 1, 1)$. Find equations for
   (i) the lines containing the edges through P,
   (ii) the lines containing the diagonals (through P) of the faces through P,
   (iii) the lines containing the diagonals (not through P) of the faces through P.

2. Find the lines through the pairs of points with the following coordinates:
   (i) $(1, 1, 1)$ and $(4, 2, 3)$,
   (ii) $(2, 0, 1)$ and $(7, 4, 2)$,
   (iii) $(1, 4, 2)$ and $(0, 5, 3)$.

3. Find the angle between any two of the lines in Ex. 2.

4. In a regular tetrahedron $ABCD$ find the cosine of the angle between the edge AD and the face ABC. Find also the angle between two faces. (The angle between a line and a plane is defined as the complement of the acute angle between the line and the normal to the plane; the angle between two planes is defined as the acute angle between their normals.)

5. Find the planes through the following sets of points:
   (i) $(1, 2, 3)$, $(3, 5, 7)$, $(3, -1, -3)$,
   (ii) $(4, -5, 8)$, $(1, 2, 3)$, $(2, 9, -4)$,
   (iii) $(0, 12, 2)$, $(2, -2, 14)$, $(1, 2, 3)$.

6. Find the intersections of any two of the planes in Ex. 5.

7. Find the points where the common perpendicular to each of the following pairs of lines meets these lines:
   (i) $(1, 4, 0)+\lambda(0, 1, 1)$ and $(1, -1, 4)+\mu(2, 1, 2)$,
   (ii) $(0, 1, 1)+\lambda(1, 1, 1)$ and $(1, 0, 0)+\mu(1, 2, -3)$.

8. If **a**, **b**, **c** are any vectors show that the four points with position-vectors **a**, **b**, **c**, $\frac{1}{3}(\mathbf{a}+\mathbf{b}+\mathbf{c})$ are coplanar.

9. The position-vectors of three vertices of a tetrahedron relative to the fourth as origin are **a**, **b** and **c**. Through each vertex a plane is drawn parallel to the opposite face. Express the position-vectors of the vertices of the tetrahedron formed by these planes in terms of **a**, **b** and **c**.

CHAPTER THREE

Coordinate Transformations

1. We have seen that the points of space may be described as linear combinations of three linearly independent vectors, the scalar coefficients being the coordinates relative to these basis-vectors (1.6). Our main object in this chapter is to compare the coordinates of a given point relative to two sets of basis-vectors. Let $\mathbf{e}_1$, $\mathbf{e}_2$, $\mathbf{e}_3$ and $\mathbf{f}_1$, $\mathbf{f}_2$, $\mathbf{f}_3$ be any two sets of basis-vectors and consider the two coordinate-systems defined by these sets of vectors, relative to a given point O as origin (the same for both systems). For a given point P we have

$$OP=\sum \mathbf{e}_i x_i \tag{1}$$

and

$$OP=\sum \mathbf{f}_i y_i, \tag{2}$$

say. To obtain a relation between the x's and y's, let us express the vectors $\mathbf{f}_1$, $\mathbf{f}_2$, $\mathbf{f}_3$ in terms of the $\mathbf{e}$'s. We have

$$\mathbf{f}_j=\sum \mathbf{e}_i a_{ij} \quad (j=1, 2, 3) \tag{3}$$

and inserting this expression in (2), we find

$$OP=\sum \mathbf{e}_i a_{ij} y_j, \tag{4}$$

where the summation is over i and j. Now the coordinates of P relative to $\mathbf{e}_1$, $\mathbf{e}_2$, $\mathbf{e}_3$ are uniquely determined; we may therefore equate the coefficients of $\mathbf{e}_1$ in (1) and (4) and obtain

$$x_1=\sum a_{1j} y_j$$

and similarly with 1 replaced by 2 and 3. Thus

$$x_i=\sum a_{ij} y_j. \quad (i=1, 2, 3). \tag{5}$$

In matrix notation this may be written†

† cf. LE, p. 28.

$$x=Ay, \tag{6}$$

where x and y are the columns consisting of x_1, x_2, x_3 and y_1, y_2, y_3 respectively and A is the 3×3 matrix whose general element is a_{ij}. We note that by (3) the columns of A represent the coordinates of $\mathbf{f}_1$, $\mathbf{f}_2$, $\mathbf{f}_3$ relative to the **e**'s as basis-vectors.

Of course the same argument may be carried out with the roles of the **e**'s and **f**'s interchanged. We have to express $\mathbf{e}_1$, $\mathbf{e}_2$, $\mathbf{e}_3$ in terms of the **f**'s,

$$\mathbf{e}_j=\sum \mathbf{f}_i b_{ij} \quad (i=1, 2, 3) \tag{7}$$

say, and from this equation together with (2) we find

$$y_i=\sum b_{ij}x_j$$

or in matrix form

$$y=Bx, \tag{8}$$

where $B=(b_{ij})$. Equations (6) and (8) express the transformation of coordinates from x to y and vice versa. If we substitute from (8) into (6) we find

$$x=ABx$$

and since this holds for all columns x, we must have† $AB=I$, and similarly $BA=I$. Thus

$$B=A^{-1}; \tag{9}$$

in particular this shows that the matrix A of the transformation is regular.§

Conversely, given a coordinate-system with basis-vectors $\mathbf{e}_1$, $\mathbf{e}_2$, $\mathbf{e}_3$ say, any regular matrix A defines a coordinate transformation. We need only define three vectors $\mathbf{f}_1$, $\mathbf{f}_2$, $\mathbf{f}_3$ by the rule

$$\mathbf{f}_j \leftrightarrow (a_{1j}, a_{2j}, a_{3j}). \tag{10}$$

Since the columns of A are linearly independent, so are the vectors $\mathbf{f}_1$, $\mathbf{f}_2$, $\mathbf{f}_3$ and hence they define again a coordinate-system. Since the coordinates of $\mathbf{f}_j$ constitute a column in the matrix A, we shall often write (10) as

$$\mathbf{f}_j \leftrightarrow a_j \quad (j=1, 2, 3)$$

† Here I is the unit-matrix. § cf. LE, p. 39.

where a_j is the j-th column of A.

2. We now take two *rectangular* coordinate-systems (still with the same origin) and ask for the conditions which the matrix of transformation has to satisfy. If this matrix is again denoted by A, we have to express the fact that the columns of A represent mutually orthogonal unit-vectors (in a rectangular system). Thus

$$a_i' a_j = \delta_{ij} \quad (i, j = 1, 2, 3),$$

where a_i' denotes the transpose of a_i and δ_{ij} is again the Kronecker delta (cf. p. 12).

These equations may be summed up in a single matrix equation

$$A'A = I. \tag{11}$$

Conversely, for a given rectangular system the transformation defined by a matrix satisfying (11) leads again to a rectangular system.

Any square matrix A satisfying (11) is said to be *orthogonal*, and the corresponding transformation is called an *orthogonal transformation.* We note that for any orthogonal matrix A,

$$1 = |A'A| = |A'|\,|A| = |A|^2,$$

and hence†

$$|A| = \pm 1. \tag{12}$$

The matrix A (as well as the transformation defined by it) is called *proper* or *improper orthogonal* according as $|A|$ is $+1$ or -1. The geometrical interpretation will be given later (No. 3).

An orthogonal matrix A necessarily has an inverse (since its determinant is not zero, by (12)), and as (11) shows, the inverse is its transpose:

$$A^{-1} = A'. \tag{13}$$

It follows that the inverse of an orthogonal matrix A is again

† Of course this notation for the determinant of a matrix must not be confused with the absolute value of a real number, defined on p. 4.

orthogonal, for this inverse is A', and we have $A''A' = AA' = AA^{-1} = I$. The product of two orthogonal matrices is also orthogonal, for if $A'A = B'B = I$, then

$$(AB)'AB = B'A'AB = B'B = I.$$

Moreover, if A and B are proper orthogonal, then so is their product, because $|AB| = |A|\,|B| = 1$.

Equation (13) greatly simplifies the discussion of orthogonal transformations. Thus the transformation from one rectangular system to another was described by the equation (6):

$$x = Ay,$$

where A is orthogonal. If we want to express the new coordinates y in terms of the old, we need the inverse of A; but this is A' as we saw in (13). Hence the inverse transformation is

$$y = A'x. \qquad (14)$$

For example, if the equations of transformation are

$$x = \begin{pmatrix} \frac{1}{\sqrt{3}} & \frac{1}{\sqrt{2}} & -\frac{1}{\sqrt{6}} \\ \frac{1}{\sqrt{3}} & -\frac{1}{\sqrt{2}} & -\frac{1}{\sqrt{6}} \\ -\frac{1}{\sqrt{3}} & 0 & -\frac{2}{\sqrt{6}} \end{pmatrix} y,$$

then the inverse transformation is given by

$$y = \begin{pmatrix} \frac{1}{\sqrt{3}} & \frac{1}{\sqrt{3}} & -\frac{1}{\sqrt{3}} \\ \frac{1}{\sqrt{2}} & -\frac{1}{\sqrt{2}} & 0 \\ -\frac{1}{\sqrt{6}} & -\frac{1}{\sqrt{6}} & -\frac{2}{\sqrt{6}} \end{pmatrix} x.$$

If we have a third coordinate-system, again rectangular and with the same origin as the other two, in which the coordinates of the general point P are denoted by z, then this is related to the second system by an orthogonal transformation, say

$$y=Bz. \tag{15}$$

The transformation from the original system to the final system is obtained by eliminating y from the equations (6) and (15),

$$x=ABz. \tag{16}$$

Since (16) and (14) are transformations between rectangular coordinate-systems, these equations provide another proof of the fact that products and inverses of orthogonal matrices are again orthogonal.

3. We now consider some examples of orthogonal transformations. If we rotate the axes of a rectangular coordinate-system like a rigid body about the origin, we shall again obtain a rectangular system, and the resulting transformation must therefore be orthogonal. In order to determine the matrix of a rotation we need to describe the coordinate-systems more closely.

Suppose we have two rectangular coordinate-systems with the same origin O. If we think of each system as rigidly movable about O, we can rotate the second system so that its 1-axis coincides (as oriented line) with the 1-axis of the first system. Now the second system may be rotated about the common 1-axis so that their 2-axes coincide. Then the 3-axes of the systems lie along the same line but they either have the same orientation or have opposite orientations.

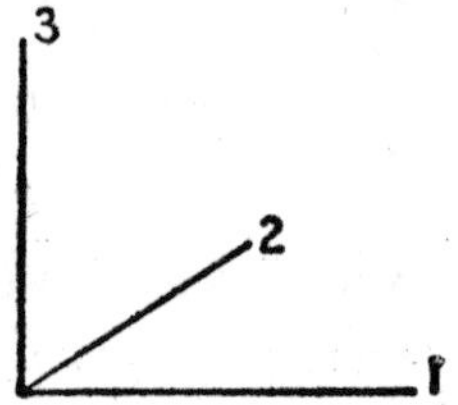

FIG. 9

Accordingly we can distinguish two possible *orientations* of a rectangular coordinate-system: If the rotation through 90° from the 1-axis to the 2-axis is related to the positive direction of the 3-axis as are rotation and motion of a right-handed screw, the system is said to be *right-handed* (Fig. 9), in the opposite case it is *left-handed* (Fig. 10). The distinc-

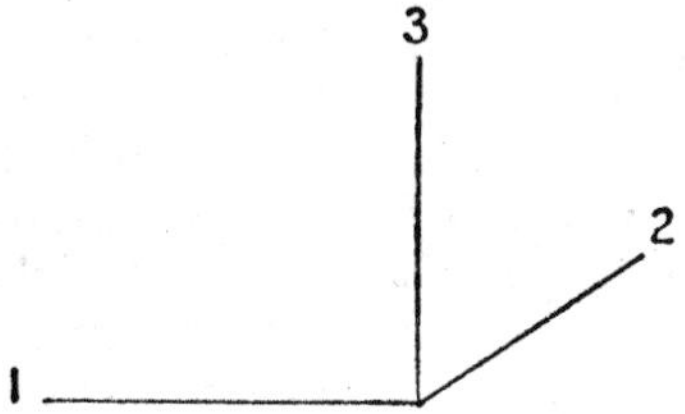

FIG. 10

tion may easily be remembered by the fact that thumb, index and middle finger of the right hand when extended, form a right-handed system, while the left hand exemplifies a left-handed system. In this way the orientation may be defined for any system, not necessarily rectangular. It is clear that a rotation can never change a right-handed system into a left-handed one, nor vice versa. In other words, the orientation of a coordinate-system is preserved by rotations.

Consider now the matrix of transformation when a right-handed rectangular system is rotated about the 3-axis. Denote the angle of rotation by α (the sense of rotation being related to the positive direction on the axis as in a right-handed screw); then the basis-vectors of the new system, expressed in terms of the old, have the coordinates

$$(\cos\alpha, \sin\alpha, 0), \ (-\sin\alpha, \cos\alpha, 0) \text{ and } (0, 0, 1).$$

These are therefore the columns of the required matrix, and the transformation is given by

$$x = \begin{pmatrix} \cos\alpha & -\sin\alpha & 0 \\ \sin\alpha & \cos\alpha & 0 \\ 0 & 0 & 1 \end{pmatrix} y,$$

with a proper orthogonal matrix. Written out in terms of scalar equations the transformation has the form

$$x_1 = y_1 \cos\alpha - y_2 \sin\alpha$$
$$x_2 = y_1 \sin\alpha + y_2 \cos\alpha$$
$$x_3 = y_3.$$

In a similar way it may be shown that any rotation is represented by a proper orthogonal matrix. Given two right-handed rectangular systems (with the same origin), we can always get from one to the other by a series of rotations (cf. the argument on p. 38), and the transformation from one to the other is therefore proper orthogonal. On the other hand, the transformation from a right-handed to a left-handed system must be improper. For if x and y denote the coordinates of a given point in the two systems, we introduce a third system whose 1- and 2-axes agree with those of the second system, but whose 3-axis has the opposite orientation. This system is therefore right-handed; if the same point has coordinates z in this system, then

$$x = Az,$$

where A, as matrix of transformation between right-handed systems, is proper orthogonal. On the other hand, $z_1 = y_1$, $z_2 = y_2$, $z_3 = -y_3$, i.e.

$$z = Ty,$$

where T, given by the equation

$$T = \begin{pmatrix} 1 & 0 & 0 \\ 0 & 1 & 0 \\ 0 & 0 & -1 \end{pmatrix} \tag{17}$$

is an improper orthogonal matrix. The transformation from x to y is therefore

$$x = ATy,$$

and $|AT| = |A|\,|T| = -1$, i.e. the transformation is improper. The transformation described by (17) consists in reflecting every point in the 12-plane, as in a mirror. Such a transformation is called a *reflexion*, and it may be shown that any two coordinate-systems related by a reflexion have opposite orientations.

4. With any triad of vectors $\mathbf{a}_1$, $\mathbf{a}_2$, $\mathbf{a}_3$, a scalar called their *triple product* may be associated as follows: Let

$$\mathbf{a}_i \leftrightarrow x_i$$

in some right-handed rectangular coordinate-system. Then the triple product of $\mathbf{a}_1$, $\mathbf{a}_2$, $\mathbf{a}_3$ is defined as the determinant of the matrix with x_1, x_2, x_3 as columns:

$$[\mathbf{a}_1, \mathbf{a}_2, \mathbf{a}_3] = |x_1, x_2, x_3|. \tag{18}$$

If in a general coordinate-system we have

$$\mathbf{a}_i \leftrightarrow y_i,$$

then

$$x_i = Ay_i \quad (i=1, 2, 3),$$

with a regular matrix A of transformation. Hence

$$\begin{aligned} |x_1, x_2, x_3| &= |Ay_1, Ay_2, Ay_3| \\ &= |A|\,|y_1, y_2, y_3|; \end{aligned}$$

i.e.

$$|x_1, x_2, x_3| = |A|\,|y_1, y_2, y_3|. \tag{19}$$

We consider especially the case where the y-system is again rectangular. Then A is an orthogonal matrix, proper or improper according as the y-system is right- or left-handed. According to (19) we therefore have

$$[\mathbf{a}_1, \mathbf{a}_2, \mathbf{a}_3] = |y_1, y_2, y_3| \text{ in a right-handed system,} \tag{20}$$

$$[\mathbf{a}_1, \mathbf{a}_2, \mathbf{a}_3] = -|y_1, y_2, y_3| \text{ in a left-handed system.} \tag{21}$$

Of course (19) may also be used to express the triple product in oblique coordinates, although we shall not have occasion to do so.

We can also use the triple product to give an invariant definition of the vector product. Let $\mathbf{a}$ and $\mathbf{b}$ be any vectors and $\mathbf{a} \wedge \mathbf{b}$ their vector product in a fixed right-handed rectangular system, as defined by (25) of Chapter 1. Then for any vector $\mathbf{c}$, we have, by the expansion rule of determinants (LE, p. 59),

$$[\mathbf{a}, \mathbf{b}, \mathbf{c}] = (\mathbf{a} \wedge \mathbf{b}) \,.\, \mathbf{c}. \tag{22}$$

Together with (20) this shows that the same rule (25) of Chapter 1 may be used to obtain the coordinates of $\mathbf{a} \wedge \mathbf{b}$

in *any* right-handed rectangular coordinate-system, while in a left-handed rectangular system we have, by (21),

$$\mathbf{a} \wedge \mathbf{b} \leftrightarrow -(a_2b_3-a_3b_2,\ a_3b_1-a_1b_3,\ a_1b_2-a_2b_1) \qquad (23)$$

in terms of the coordinates (a_1, a_2, a_3) of **a** and (b_1, b_2, b_3) of **b**.

Thus the general definition of $\mathbf{a} \wedge \mathbf{b}$, in any rectangular coordinate-system—right- or left-handed—with basis-vectors $\mathbf{e}_1, \mathbf{e}_2, \mathbf{e}_3$ may be written

$$\mathbf{a} \wedge \mathbf{b}=[\mathbf{a}, \mathbf{b}, \mathbf{e}_1]\mathbf{e}_1+[\mathbf{a}, \mathbf{b}, \mathbf{e}_2]\mathbf{e}_2+[\mathbf{a}, \mathbf{b}, \mathbf{e}_3]\mathbf{e}_3. \qquad (24)$$

To obtain a geometrical interpretation of the triple product we use (22) and note that $\mathbf{a} \wedge \mathbf{b}$ is a vector at right angles to both **a** and **b**, and of length equal to the area of the parallelogram spanned by **a** and **b**. Now [**a**, **b**, **c**] is the product of this area by the projection of **c** on the perpendicular to **a** and **b**, in other words [**a**, **b**, **c**] is just the volume of the parallelepiped spanned by **a**, **b**, **c**, this volume being base × height. This shows also that [**a**, **b**, **c**] vanishes if and only if **a**, **b**, **c** lie in a plane, i.e. are linearly dependent; of course this also follows from the determinantal criterion for linear dependence (LE, p. 68). Finally, to determine the sign we choose a rectangular system with basis-vectors $\mathbf{e}_1, \mathbf{e}_2, \mathbf{e}_3$ so that $\mathbf{e}_1$ is a positive scalar multiple of **a**, $\mathbf{e}_2$ is in the plane of **a** and **b**, such that $\mathbf{b} \cdot \mathbf{e}_2>0$ and $\mathbf{e}_3$ is such that $\mathbf{c} \cdot \mathbf{e}_3>0$. This is always possible if **a**, **b**, **c** are linearly independent, and by construction $\mathbf{e}_1, \mathbf{e}_2, \mathbf{e}_3$ forms a right-handed or left-handed system according as the set **a**, **b**, **c** is right-handed or left-handed. Moreover, in this system, $\mathbf{a}\leftrightarrow(a_1, 0, 0)$, $\mathbf{b}\leftrightarrow(b_1, b_2, 0)$, $\mathbf{c}\leftrightarrow(c_1, c_2, c_3)$ and a_1, b_2, c_3 are all positive. Hence the determinant of the coordinates is positive and by (20) and (21), [**a**, **b**, **c**] is positive or negative or zero according as **a**, **b**, **c** is a right-handed set or a left-handed set or is linearly dependent.

If we take $\mathbf{c}=\mathbf{a} \wedge \mathbf{b}$ in (22) we see from the last result that $\mathbf{a} \wedge \mathbf{b}$ is oriented so that **a**, **b**, $\mathbf{a} \wedge \mathbf{b}$ form a right-handed set, unless **a** and **b** are linearly dependent.

As an example of the application of vector products and

triple products we calculate the shortest distance between two skew lines. Let the lines be l and m, given by equations

$$\mathbf{x}=\mathbf{p}+\lambda\mathbf{a} \quad (\mathbf{a}\neq\mathbf{o}),$$
$$\mathbf{x}=\mathbf{q}+\mu\mathbf{b} \quad (\mathbf{b}\neq\mathbf{o}).$$

The shortest distance between these lines is the length of their common perpendicular, as we saw in 2.8. This may be obtained as the projection on the common perpendicular to the lines of any vector from a point on l to a point on m. Hence the shortest distance is $\mathbf{u}\,.\,(\mathbf{q}-\mathbf{p})$, where $\mathbf{u}$ is a unit-vector perpendicular to both $\mathbf{a}$ and $\mathbf{b}$. Such a unit-vector may be obtained by normalizing $\mathbf{a}\wedge\mathbf{b}$, and so we obtain the expression

$$[\mathbf{a}, \mathbf{b}, \mathbf{q}-\mathbf{p}]/|\mathbf{a}\wedge\mathbf{b}|$$

for the shortest distance. In particular, the lines l and m intersect if and only if

$$[\mathbf{a}, \mathbf{b}, \mathbf{q}-\mathbf{p}]=0. \tag{25}$$

In the example on p. 28, $\mathbf{a}\wedge\mathbf{b}\leftrightarrow(5, 10, 15)$ and $\mathbf{q}-\mathbf{p}\leftrightarrow(9, -7, -3)$, hence the shortest distance is $14/\sqrt{14}=\sqrt{14}\doteqdot 3.74$.

5. We have seen that any rotation is represented by a proper orthogonal transformation. Conversely we now show that every proper orthogonal transformation represents a rotation, and at the same time determine the nature of improper transformations.

Theorem. *Let A be a 3×3 orthogonal matrix. If A is proper, it represents a rotation about some axis through the origin; if A is improper it represents a rotation, followed by a reflexion in a plane through the origin.*

To prove this theorem, let A be a proper orthogonal matrix. Suppose first that A is a 2×2 matrix, say

$$A=\begin{pmatrix} a_1 & b_1 \\ a_2 & b_2 \end{pmatrix}.$$

Since the columns represent unit-vectors, we have

$a_1^2+a_2^2=b_1^2+b_2^2=1$; if we regard (a_1, a_2) and (b_1, b_2) as plane coordinates then in terms of polar coordinates, we have

$$A=\begin{pmatrix} \cos\alpha & \cos\beta \\ \sin\alpha & \sin\beta \end{pmatrix}. \qquad (26)$$

Now $|A|=1$, therefore $\sin(\beta-\alpha)=\sin\beta\cos\alpha-\cos\beta\sin\alpha =1$, and so $\beta=\alpha+90°+n.360°$ (where n is an integer). Inserting this value in (26), we obtain

$$A=\begin{pmatrix} \cos\alpha & -\sin\alpha \\ \sin\alpha & \cos\alpha \end{pmatrix},$$

which represents a rotation through an angle α in the plane.

Now let A be a 3×3 proper orthogonal matrix. We first show that $A-I$ is singular. For we have

$$A-I=A-A'A=(I-A')A.$$

Taking determinants, we find that

$$|A-I|=|I-A'|\,|A|.$$

Now $|A|=1$ by hypothesis, and so

$$\begin{aligned}|A-I|&=|I-A'|\\ &=|I-A|,\end{aligned}$$

by transposition. But we also have

$$|I-A|=(-1)^3\,|A-I|=-|A-I|,$$

therefore $|A-I|=-|A-I|$, i.e.

$$|A-I|=0. \qquad (27)$$

We note as a consequence of (27) that the equation

$$Av=v \qquad (28)$$

has a non-zero solution v. For this is a homogeneous system of equations with vanishing determinant (cf. LE, pp. 42, 66).

If A represents the transformation between two right-handed rectangular coordinate-systems, we have an equation

$$x=Ay$$

relating the coordinates of a given point in the two systems. Now the equation (28) means that there is a line through O whose points have the same coordinates in both systems. We take this line as the 3-axis of a right-handed rectangular coordinate-system. If the coordinates of its basis-vectors, referred to the x-system, are u_1, u_2, u_3 respectively, then u_3 is a solution of (28), so that $Au_3=u_3$. Moreover, since A is proper orthogonal, Au_1, Au_2, Au_3 again form a right-handed rectangular system. The two sets u_1, u_2, u_3 and Au_1, Au_2, Au_3 have the third vector in common, and hence Au_1, Au_2 are related to u_1, u_2 by a 2×2 proper orthogonal matrix; in other words A represents a rotation about the line

$$\mathbf{x} \leftrightarrow \lambda u_3.$$

If A is improper, we take any reflexion, e.g. T, given by (17), then $R=AT$ is a proper orthogonal matrix. By what has been proved, R is a rotation, and moreover, since $T^2=I$,

$$A=RT. \tag{29}$$

Thus A has been expressed as the product of a rotation and a reflexion and the proof is complete.

We note that in the representation (28) of an improper orthogonal matrix the reflexion can be chosen arbitrarily and the rotation is then uniquely determined. Similarly, every improper orthogonal matrix can be written as the product of an (arbitrarily chosen) reflexion and a rotation, but the rotation cannot be omitted altogether, as e.g. in the matrix

$$\begin{pmatrix} -1 & 0 & 0 \\ 0 & -1 & 0 \\ 0 & 0 & -1 \end{pmatrix}$$

which is improper orthogonal, but not itself a reflexion.

It can be shown that every rotation in space can be written as the product of two reflexions. Therefore every orthogonal 3×3 matrix can be expressed as the product of at most three reflexions.

EXERCISES ON CHAPTER THREE

1. A right-handed rectangular coordinate-system is rotated through an angle of 120° about the line $\mathbf{x} \leftrightarrow \lambda(1, 1, 1)$; find the transformation. Verify that its matrix A is proper orthogonal and show also that $A^3 = I$.

2. Let $\mathbf{a}$ be any non-zero vector. Verify that the transformation defined by
$$\mathbf{x} = \mathbf{y} - 2\frac{\mathbf{a} \cdot \mathbf{y}}{\mathbf{a} \cdot \mathbf{a}}\mathbf{a}$$
is orthogonal, and give a geometrical interpretation.

3. A matrix S is said to be *skew-symmetric* if it equals its negative transpose, i.e. if $S = -S'$. Show that any orthogonal matrix A such that $I + A$ is regular can be expressed in the form
$$A = (I - S)(I + S)^{-1},$$
where S is a skew-symmetric matrix.
[Hint: Show that S can be found to satisfy
$$(I + A)(I + S) = (I + S)(I + A) = 2I.]$$

4. If A is an orthogonal matrix such that $I + A$ is singular, then either A is improper or A represents a rotation through 180°.

5. Show that three points A, B, C with position-vectors $\mathbf{a}$, $\mathbf{b}$, $\mathbf{c}$ respectively are coplanar with the origin if and only if $[\mathbf{a}, \mathbf{b}, \mathbf{c}] = 0$.

6. Show that the equation of the plane through three non-collinear points with position-vectors $\mathbf{a}$, $\mathbf{b}$, $\mathbf{c}$ may be written
$$[\mathbf{x} - \mathbf{a}, \mathbf{x} - \mathbf{b}, \mathbf{x} - \mathbf{c}] = 0.$$
[Hint: Use Ex. 5 to express the condition for four points to be coplanar.]

7. In any coordinate-system with basis-vectors $\mathbf{a}_1, \mathbf{a}_2, \mathbf{a}_3$, let $\mathbf{b}_1, \mathbf{b}_2, \mathbf{b}_3$ be such that $\mathbf{a}_i \cdot \mathbf{b}_j = \delta_{ij}$ (cf. 1, Ex. 6). Show that the coordinates of $\mathbf{x} \wedge \mathbf{y}$ in this system are $[\mathbf{x}, \mathbf{y}, \mathbf{b}_1]$, $[\mathbf{x}, \mathbf{y}, \mathbf{b}_2]$, $[\mathbf{x}, \mathbf{y}, \mathbf{b}_3]$, thus
$$\mathbf{x} \wedge \mathbf{y} = \Sigma[\mathbf{x}, \mathbf{y}, \mathbf{b}_i]\, \mathbf{a}_i.$$

8. For any vectors $\mathbf{a}$, $\mathbf{b}$, $\mathbf{c}$ show that
$$(\mathbf{a} \wedge \mathbf{b}) \wedge \mathbf{c} = (\mathbf{a} \cdot \mathbf{c})\mathbf{b} - (\mathbf{b} \cdot \mathbf{c})\mathbf{a}.$$
[Hint: Show first that the right-hand side is linearly dependent on $\mathbf{a}$ and $\mathbf{b}$ and determine the ratio of the coefficients by taking scalar products with $\mathbf{c}$. This determines the right-hand side up to a constant factor which may be found e.g. by comparing coefficients of $a_1 b_2 c_1$ in a rectangular coordinate-system.]

9. For any vectors $\mathbf{a}$, $\mathbf{b}$, $\mathbf{c}$ show that
$$(\mathbf{a} \wedge \mathbf{b}) \wedge \mathbf{c} + (\mathbf{b} \wedge \mathbf{c}) \wedge \mathbf{a} + (\mathbf{c} \wedge \mathbf{a}) \wedge \mathbf{b} = 0.$$
[Hint: Use Ex. 8.]

10. For any vectors $\mathbf{a}$, $\mathbf{b}$, $\mathbf{c}$, $\mathbf{d}$ show that
$$(\mathbf{a} \wedge \mathbf{b}) \cdot (\mathbf{c} \wedge \mathbf{d}) = (\mathbf{a} \cdot \mathbf{c})(\mathbf{b} \cdot \mathbf{d}) - (\mathbf{a} \cdot \mathbf{d})(\mathbf{b} \cdot \mathbf{c}),$$
and apply the result to prove (29) of Ch. 1.

11. Show that if $\mathbf{a}$ is any vector and $\mathbf{u}$ a unit-vector orthogonal to $\mathbf{a}$, then the equation

$$\mathbf{u} \wedge \mathbf{x} = \mathbf{a}$$

represents the line in the direction $\mathbf{u}$ through the point $\mathbf{u} \wedge \mathbf{a}$.

12. Find the shortest distance between pairs of opposite edges (produced if necessary) of the tetrahedron $ABCD$, where

$$A \leftrightarrow (0, 0, 0),\ B \leftrightarrow (3, 1, 2),\ C \leftrightarrow (2, -1, 4),\ D \leftrightarrow (-1, 3, 2)$$

(in some rectangular coordinate-system).

CHAPTER FOUR

Spheres

1. The sphere is the simplest type of surface next to the plane. It is defined as the set of points in space at a fixed distance from a given point. The fixed distance is called the *radius* and the given point the *centre* of the sphere.

It is an easy matter to write down the equation of the general sphere. Let $\mathbf{a}$ be the position-vector of the centre and r the radius of the sphere. The condition for the point with position-vector $\mathbf{x}$ to lie on the sphere is that the difference $\mathbf{x}-\mathbf{a}$ be of length r, i.e.

$$|\mathbf{x}-\mathbf{a}|=r. \qquad (1)$$

This is therefore the equation of the sphere of radius r and centre $\mathbf{a}$. Working out the left-hand side, we may also write the equation as

$$|\mathbf{x}|^2-2\mathbf{a}\,.\,\mathbf{x}+|\mathbf{a}|^2=r^2. \qquad (2)$$

This is of the form

$$|\mathbf{x}|^2-2\mathbf{a}\,.\,\mathbf{x}+\rho=0, \qquad (3)$$

where $\mathbf{a}$ is a vector and ρ is a scalar such that

$$|\mathbf{a}|^2-\rho>0. \qquad (4)$$

Conversely, if $\mathbf{a}$ and ρ satisfy (4), then (3) represents a sphere, for it is equivalent to (2), where r is the positive square root of $|\mathbf{a}|^2-\rho$.

More generally, the equation

$$\lambda|\mathbf{x}|^2-2\mathbf{a}\,.\,\mathbf{x}+\rho=0 \qquad (5)$$

represents a sphere if and only if

$$\lambda\neq 0 \qquad (6)$$

and

$$|\mathbf{a}|^2-\lambda\rho>0. \qquad (7)$$

For if (6) holds, then we may divide (5) by λ to obtain an equation of the form (3), with $\frac{1}{\lambda}\mathbf{a}$, $\frac{1}{\lambda}\rho$ in place of $\mathbf{a}$, ρ. The condition (4) for this equation to represent a sphere is therefore given by (7). On the other hand if (6) is not satisfied, then $\lambda=0$ and (5) reduces to a plane or (if $\mathbf{a}=\mathbf{o}$) represents no point at all.

2. In order to determine the intersection of a sphere and a plane let us take a rectangular coordinate-system whose origin and first two basis-vectors lie in the given plane. Thus the plane forms the 12-plane in this coordinate-system.

Then (2) may be rewritten

$$x_1^2+x_2^2+x_3^2-2(a_1x_1+a_2x_2+a_3x_3)=r^2-|\mathbf{a}|^2,$$

where $\mathbf{x}\leftrightarrow(x_1, x_2, x_3)$ and $\mathbf{a}\leftrightarrow(a_1, a_2, a_3)$. The intersection with the plane is found by putting $x_3=0$:

$$x_1^2+x_2^2-2(a_1x_1+a_2x_2)=r^2-|\mathbf{a}|^2,$$

which may also be written

$$(x_1-a_1)^2+(x_2-a_2)^2=r^2-a_3^2.$$

This equation represents a circle in the 12-plane provided that $r^2-a_3^2>0$. If $r^2=a_3^2$, it represents the single point $(a_1, a_2, 0)$, while if $r^2<a_3^2$, no point of the 12-plane satisfies the equation.

We see therefore that a sphere and a plane intersect either in a circle or in a point or not at all. Moreover, the intersection, when it exists, is determined completely when its centre and radius are given. They may be found as follows, if we note that the centre of the circle of intersection is the foot of the perpendicular from the centre of the sphere to the plane.† If the sphere is given by (2) and the plane by

$$\mathbf{x}\,.\,\mathbf{t}=p, \tag{8}$$

† This is easily seen from geometrical considerations, or may be verified by using the special coordinate-system chosen above.

then this perpendicular is

$$\mathbf{x}=\mathbf{a}+\lambda\mathbf{t},$$

and the point of intersection with the plane is given by $\lambda=\lambda_0$, where λ_0 is the solution of

$$(\mathbf{a}+\lambda\mathbf{t})\,.\,\mathbf{t}=p,$$

i.e.

$$\lambda_0=(p-\mathbf{a}\,.\,\mathbf{t})/|\mathbf{t}|^2.$$

The position-vector of the centre of the circle is therefore

$$\mathbf{a}+(p-\mathbf{a}\,.\,\mathbf{t})\mathbf{t}/|\mathbf{t}|^2. \tag{9}$$

The distance of the centre of the sphere from the plane is $|\lambda_0\mathbf{t}|=|p-\mathbf{a}\,.\,\mathbf{t}|/|\mathbf{t}|$, and the radius of the circle is r_1, where

$$r_1^2=r^2-(p-\mathbf{a}\,.\,\mathbf{t})^2/|\mathbf{t}|^2. \tag{10}$$

We note that the second term on the right represents the square of the distance of the plane from the centre of the sphere. E.g., to find the intersection of the sphere

$$x_1^2+x_2^2+x_3^2-2x_3=0$$

and the plane

$$x_1+x_2+x_3=1,$$

we have $\mathbf{a}\leftrightarrow(0, 0, 1)$, $\mathbf{t}\leftrightarrow(1, 1, 1)$, $r=1$, $p=1$, hence $p-\mathbf{a}\,.\,\mathbf{t}=0$ and so the centre of the intersection is $\mathbf{a}\leftrightarrow(0, 0, 1)$, while the radius is $r_1=1$. Thus the plane passes through the centre of the sphere.

3. In order to obtain the intersection of two spheres we shall find a plane passing through their intersection, so that the result of No. 2 can be applied to find the intersection. Let the spheres be

$$|\mathbf{x}|^2-2\mathbf{a}\,.\,\mathbf{x}+\rho=0, \tag{11}$$

$$|\mathbf{x}|^2-2\mathbf{b}\,.\,\mathbf{x}+\sigma=0. \tag{12}$$

If these spheres are concentric, they cannot intersect unless they coincide. Excluding this case, we may assume $\mathbf{a}\neq\mathbf{b}$; then the equation obtained by subtracting (12) from (11),

$$2(\mathbf{b}-\mathbf{a})\,.\,\mathbf{x}+\rho-\sigma=0 \tag{13}$$

represents a plane. Moreover, any point satisfying two of the equations (11), (12), (13) must also satisfy the third, so that the plane (13) meets each of the spheres (11), (12) in their common part. The plane (13) is called the *radical plane* of the spheres (11) and (12); we note that it is defined whether these planes intersect or not (so long as they are not concentric).

4. It is clear geometrically, that a given plane intersects a sphere of radius r if and only if its distance from the centre of the sphere does not exceed r. If its distance is exactly r, then by (10), the intersection is a single point ('circle of radius zero') and the plane is then the tangent plane at this point.

To find the equation of the tangent plane to the sphere (2) at the point P of the sphere we have to express the fact that the plane passes through P and is perpendicular to the vector from the centre to P. If $\mathbf{b}$ is the position vector of P, then the equation of the tangent plane is

$$(\mathbf{x}-\mathbf{b}) \cdot (\mathbf{a}-\mathbf{b})=0,$$

or

$$(\mathbf{a}-\mathbf{b}) \cdot \mathbf{x}=(\mathbf{a}-\mathbf{b}) \cdot \mathbf{b}.$$

5. Through four points there passes in general just one sphere. To obtain its equation we consider five general points, A, B, C, D, E and express the condition that they lie on a sphere. If the position-vectors of the points are $\mathbf{a}$, $\mathbf{b}$, $\mathbf{c}$, $\mathbf{d}$, $\mathbf{e}$ respectively, then the condition is that scalars λ, k_1, k_2, k_3, ρ may be found satisfying

$$\lambda \neq 0, \quad k_1^2+k_2^2+k_3^2-\lambda\rho>0,$$

such that†

$$|a|^2-2a_1k_1-2a_2k_2-2a_3k_3+\rho=0,$$

$$\cdot \quad \cdot \quad \cdot \quad \cdot \quad \cdot \quad \cdot$$

$$|e|^2-2e_1k_1-2e_2k_2-2e_3k_3+\rho=0,$$

where $\mathbf{a}\leftrightarrow(a_1, a_2, a_3), \ldots, \mathbf{e}\leftrightarrow(e_1, e_2, e_3)$.

† Here and in what follows, for any set of coordinates $a=(a_1, a_2, a_3)$, $a_1{}^2+a_2{}^2+a_3{}^2$ is written as $|a|^2$ for brevity.

These equations have a non-trivial solution if and only if

$$\begin{vmatrix} |a|^2 & a_1 & a_2 & a_3 & 1 \\ |b|^2 & b_1 & b_2 & b_3 & 1 \\ \cdot & \cdot & \cdot & \cdot & \cdot \\ |e|^2 & e_1 & e_2 & e_3 & 1 \end{vmatrix} = 0. \qquad (14)$$

Thus the equation (14) is necessary for the points A, B, C, D, E to lie on a sphere. Conversely, if (14) holds, we can find λ, $\mathbf{k} \leftrightarrow (k_1, k_2, k_3)$, ρ, not all zero, such that $\mathbf{a}, \ldots, \mathbf{e}$ all satisfy

$$\lambda|\mathbf{x}|^2 - 2\mathbf{k} \, . \, \mathbf{x} + \rho = 0. \qquad (15)$$

We already know that if $\lambda \neq 0$ and $|\mathbf{k}|^2 - \lambda\rho > 0$, then (15) represents a sphere. Suppose this is not so; if $\lambda \neq 0$, we must have $|\mathbf{k}|^2 - \lambda\rho \leqslant 0$ and (15) represents either (i) a single point or (ii) no point. On the other hand, if $\lambda = 0$, (15) represents either (iii) a plane or (iv) no point at all. Since the equation is satisfied by $A, \ldots, E$, cases (ii) and (iv) cannot occur, and moreover, if $A, \ldots, E$ are not coplanar, cases (i) and (iii) are excluded too and we can say:

Five points $A, \ldots, E$ which are not coplanar, lie on a sphere if and only if their coordinates satisfy the equation (14).

Now let A, B, C, D be four points which are not coplanar, then the point P with position-vector $\mathbf{x} \leftrightarrow (x_1, x_2, x_3)$ lies on a sphere with A, B, C, D if and only if

$$\begin{vmatrix} |x|^2 & x_1 & x_2 & x_3 & 1 \\ |a|^2 & a_1 & a_2 & a_3 & 1 \\ \cdot & \cdot & \cdot & \cdot & \cdot \\ |d|^2 & d_1 & d_2 & d_3 & 1 \end{vmatrix} = 0. \qquad (16)$$

If we expand this determinant by the first row, we obtain the equation of a sphere, the coefficient of $|x|^2$ being $[\mathbf{a}-\mathbf{d}, \mathbf{b}-\mathbf{d}, \mathbf{c}-\mathbf{d}]$, which is not zero because the points A, B, C, D are not coplanar. This shows incidentally, that four points which are not coplanar determine a unique sphere. On the other hand, through four coplanar points there is in general no sphere.

EXERCISES ON CHAPTER FOUR

1. Find the centre and radius of the spheres
   (i) $|x|^2+6x_1-8x_3=0$,
   (ii) $3|x|^2-10x_1-8x_2-12x_3-4/3=0$.

2. Find the equation of the sphere through the points
   (i) $(1, -1, 1)$, $(0, 1, 2)$, $(2, 3, 0)$, $(5, 2, 4)$,
   (ii) $(3, 3, 4)$, $(-3, 0, 1)$, $(-1, 3, -2)$, $(4, 3, -1)$.

3. Find the sphere which contains the intersection of
   $$|x|^2-4x_1-6x_2-8x_3-30=0 \text{ and } x_1+x_2+x_3=1$$
   and passes through the point $(4, 3, 3)$.

4. Find the intersection of the line $\mathbf{x}\leftrightarrow(2, 3, 2)+\lambda(1, 0, -1)$ and the sphere $|x|^2-4x_1-6x_2+8x_3-21=0$. Find also the equations of the tangent planes at these points.

5. Let A, B be two points with position vectors **a**, **b** respectively. Express the sphere on AB as diameter in terms of **a** and **b**.

6. How many spheres of radius 1 and centre in the 12-plane can be drawn to touch both the 1-axis and the 2-axis? Give the equation of that sphere whose centre lies in the first quadrant of the 12-plane.

7. Show that four coplanar points lie on a sphere if and only if they are concyclic, and that there are then infinitely many spheres passing through the four points.

8. Show that the equation of the sphere through four (non-coplanar) points with position-vectors **a**, **b**, **c**, **d** may be written
   $$(|\mathbf{a}|^2-|\mathbf{x}|^2)\,[\mathbf{b}-\mathbf{x}, \mathbf{c}-\mathbf{x}, \mathbf{d}-\mathbf{x}]-(|\mathbf{b}|^2-|\mathbf{x}|^2)\,[\mathbf{a}-\mathbf{x}, \mathbf{c}-\mathbf{x}, \mathbf{d}-\mathbf{x}]$$
   $$+(|\mathbf{c}|^2-|\mathbf{x}|^2)\,[\mathbf{a}-\mathbf{x}, \mathbf{b}-\mathbf{x}, \mathbf{d}-\mathbf{x}]$$
   $$-(|\mathbf{d}|^2-|\mathbf{x}|^2)\,[\mathbf{a}-\mathbf{x}, \mathbf{b}-\mathbf{x}, \mathbf{c}-\mathbf{x}]=0.$$

CHAPTER FIVE

Central Quadrics

In this chapter we shall often have to write equations referred to a (right-handed rectangular) coordinate-system, rather than in vector form. Instead of speaking of 'the vector whose coordinates are a_1, a_2, a_3' we shall simply speak of 'the vector (a_1, a_2, a_3)', leaving the coordinate-system to be understood. As before we shall abbreviate the coordinates by a single letter. For example, to say that u is a unit-vector is to say that its coordinates u_1, u_2, u_3 satisfy $u_1^2+u_2^2+u_3^2=1$, or in matrix notation, $u'u=1$, where $u'=(u_1, u_2, u_3)$ and u, its transpose, is the corresponding column.

1. A *quadric* is a surface in space given by an equation of the second degree in the coordinates. The most general equation of the second degree is

$$\left.\begin{aligned} a_{11}x_1^2+a_{22}x_2^2+a_{33}x_3^2+2a_{12}x_1x_2+2a_{13}x_1x_3+2a_{23}x_2x_3 \\ +2a_1x_1+2a_2x_2+2a_3x_3+\alpha=0, \end{aligned}\right\} \qquad (1)$$

where $a_{11}, \ldots, \alpha$ are any real numbers. The numerical factor 2 has been inserted merely for convenience, to avoid fractions later on. If we define a_{21} by $a_{21}=a_{12}$, then the term $2a_{12}x_1x_2$ may also be written as $a_{12}x_1x_2+a_{21}x_2x_1$. Similarly we define a_{31} and a_{32} by the equations $a_{31}=a_{13}$ and $a_{32}=a_{23}$; then (1) may be abbreviated as

$$\sum a_{ij}x_ix_j+2\sum a_ix_i+\alpha=0, \qquad (2)$$

where the summations are over $i, j=1, 2, 3$. The equation of a quadric may be written still more concisely in matrix notation. Thus if A is the 3×3 matrix (a_{ij}) and a the column vector $(a_1, a_2, a_3)'$, then (2) becomes

$$x'Ax+2a'x+\alpha=0. \qquad (3)$$

We note that $a_{ij}=a_{ji}$ for all i and j, by definition; these conditions may be expressed in matrix form by the equation

$$A'=A. \tag{4}$$

Any square matrix satisfying (4) is said to be *symmetric*. We note that for any 3×3 matrix B, symmetric or not, $x'Bx$ is a 1×1 matrix and therefore necessarily symmetric, i.e.

$$x'Bx=(x'Bx)'=x'B'x.$$

A complete classification of quadrics would go beyond the frame of this book. We shall therefore limit ourselves to the case where the matrix A is regular. In that case the quadric (3) is called *central*, because such a quadric has a centre of symmetry. This may be determined as follows:

If we take new coordinates y with the same basis-vectors but use the point with position vector b (in the x-system) as origin, we have

$$x=y+b;$$

then the equation (3), expressed in the y-system becomes

$$y'Ay+y'Ab+b'Ay+2a'y+\beta=0,$$

where $\beta=b'Ab+2a'b+\alpha$. By the symmetry of A this may be written

$$y'Ay+2y'(Ab+a)+\beta=0. \tag{5}$$

Now the regularity of A is precisely the condition for a unique vector b to exist, satisfying

$$Ab+a=0.$$

If we choose b in this way, equation (5) reduces to

$$y'Ay+\beta=0. \tag{6}$$

This equation contains only terms of degree two or zero in the components of y, therefore if $y=c$ is a point on the quadric, then $y=-c$ also lies on the surface. For this reason the origin in the new coordinate-system is called the *centre* of our quadric; the regularity of A ensures that there exists exactly one centre. Central quadrics include the sphere for example but exclude such quadrics as a cylinder or a paraboloid.

2. In order to obtain the tangent plane at a point of a quadric, let us consider, more generally, the points of intersection with a given line, say

$$x = p + \lambda u. \tag{7}$$

The point $p+\lambda u$ lies on the quadric (3) if and only if

$$(p+\lambda u)'A(p+\lambda u)+2a'(p+\lambda u)+\alpha=0.$$

Rearranging this equation in powers of λ, we obtain

$$\lambda^2 u'Au+2\lambda u'(Ap+a)+p'Ap+2a'p+\alpha=0. \tag{8}$$

The two roots of this equation, real, coincident or complex, give the two intersections of the line (7) with the quadric (3). In particular, (8) has 0 as a root if and only if p lies on the quadric. Consider now the case where p is a point on the quadric and q is any point other than p on the tangent plane at p. Then the line joining p and q has coincident intersections with the quadric. Putting $u=q-p$ in (7), we can express this by saying that both roots of (8) are zero, i.e. p lies on the quadric and

$$(q-p)'(Ap+a)=0.$$

As q varies we obtain the tangent plane of the quadric at p, whose equation is therefore

$$(x-p)'(Ap+a)=0. \tag{9}$$

The normal to this plane at the point p is called the *normal to the quadric at* p. Its direction is given by the vector $Ap+a$, hence the equation of the normal is

$$x=p+\mu(Ap+a). \tag{10}$$

From (9) and (10) we see that the tangent plane and normal are defined at any point p of the quadric, unless

$$Ap+a=0.$$

From No. 1 we know that this equation is satisfied only by the centre of the quadric. Thus in a central quadric, every point has a tangent plane and normal, except the centre itself. As we shall see later, the centre does not usually lie on the quadric. An example where it does is the cone, whose centre is its vertex.

3. In equation (5) we see the effect of a translation of the coordinate-system, and (6) shows how the equation of a central quadric may be simplified by a suitable translation of coordinates. We now want to consider the effect of a rotation of coordinates on the equation of the quadric. We shall assume for simplicity that the quadric is central, and is referred to the centre as origin, so that its equation is

$$x'Ax+\alpha=0. \tag{11}$$

A rotation of coordinates is given by an equation of the form

$$x=Py,$$

where P is a proper orthogonal matrix. Inserting this in (11) we obtain

$$y'P'APy+\alpha=0,$$

i.e. an equation of the form

$$y'By+\alpha=0,$$

where

$$B=P'AP. \tag{12}$$

This equation gives the relation between the matrices A and B which describe the quadric in the two coordinate-systems. We note in particular that B is again symmetric:

$$B'=(P'AP)'=P'A'P=P'AP=B.$$

If we take determinants on both sides of (12), we find

$$|B|=|P|^2\,|A|=|A|.$$

since $|P|^2=1$ (cf. 3.2). Thus the determinant $|A|$ is independent of the coordinate-system; it is called an *invariant* of the quadric. More generally, if λ is a scalar indeterminate, then

$$P'(\lambda I-A)P=\lambda P'P-P'AP=\lambda I-B,$$

and hence

$$|\lambda I-B|=|P|^2\,|\lambda I-A|=|\lambda I-A|.$$

It follows that the roots of the equation

$$|\lambda I-A|=0 \tag{13}$$

are also invariants of the quadric, i.e. they remain unchanged under any transformation of coordinates. Equation (13) is called the *characteristic equation* of A and its roots are called *characteristic values*, *eigenvalues*† or *latent roots* of A. The constant term in the equation (13) is $(-1)^3 |A| = -|A|$; this is therefore the product of the latent roots.

Since A is a 3×3 matrix, the equation (13) is of degree three; in fact the highest term is easily seen to be λ^3. It has therefore three roots, real or complex.§ If, as in the case of a quadric, A is a real symmetric matrix, these roots are necessarily real, as we shall now show. Let $\lambda = \lambda_1$ be any root of (13), possibly complex. Then there is a (complex) column vector $u \neq 0$ satisfying the equation $(\lambda_1 I - A)u = 0$; thus $Au = \lambda_1 u$, and multiplying by $\bar{u}'$, the conjugate complex of the transpose of u, we obtain

$$\bar{u}'Au = \lambda_1 \bar{u}'u. \tag{14}$$

If we take the conjugate complex of the transpose we obtain

$$\bar{u}'Au = \bar{\lambda}_1 \bar{u}'u. \tag{15}$$

Subtracting (15) from (14) we find $(\lambda_1 - \bar{\lambda}_1)\bar{u}'u = 0$. Now $u \neq 0$, and so $\bar{u}'u = |u_1|^2 + |u_2|^2 + |u_3|^2 > 0$; therefore $\lambda_1 - \bar{\lambda}_1 = 0$, i.e. λ_1 must be real.

4. We shall now prove a result which will show to what extent the equation of a central quadric may be simplified by a suitable choice of axes. Our result will apply in fact to any quadric which can be expressed in the form (11); as we have seen this includes all central quadrics, but possibly other quadrics as well.

Theorem. *Let Q be any quadric given by the equation*

$$x'Ax + \alpha = 0.$$

† *Eigen* (German) means 'characteristic'.

§ This follows from the Fundamental Theorem of Algebra, which we assume here without proof.

Then in a suitable coordinate-system its equation is

$$\lambda_1 y_1^2 + \lambda_2 y_2^2 + \lambda_3 y_3^2 + \alpha = 0,$$

where λ_1, λ_2, λ_3 *are the latent roots of the matrix* A.

In order to prove this theorem it is enough to find a proper orthogonal matrix P such that $P'AP = D$ is a diagonal matrix, say

$$D = \begin{pmatrix} \delta_1 & 0 & 0 \\ 0 & \delta_2 & 0 \\ 0 & 0 & \delta_3 \end{pmatrix}.$$

For the latent roots of D are the roots of the equation

$$(\lambda - \delta_1)(\lambda - \delta_2)(\lambda - \delta_3) = 0,$$

i.e. the numbers δ_1, δ_2, δ_3. On the other hand we have seen that A and $P'AP$ have the same latent roots, so that the diagonal elements of D are necessarily the latent roots of A.

In proving the theorem we shall use the following lemma.

Lemma. *Given any real unit-vector* u, *there exist real unit-vectors* v *and* w *so that* (u, v, w) *is a proper orthogonal matrix.*

Expressed in geometrical terms this lemma states that to any unit-vector u, two others may be found which constitute with u a right-handed set of mutually orthogonal unit-vectors (cf. 3.4). It is clear how this can be done: we choose any unit-vector v at right angles to u, i.e. we take a non-zero solution of the equation

$$u \cdot x = 0,$$

normalize it, and put $u \wedge v = w$. Then w is a unit-vector at right angles to both u and v, and u, v, w form a right-handed system (cf. 3.4). It is clear that by changing the sign of w (or by interchanging v and w) we can also obtain an *im*proper orthogonal matrix with u as first column.

We can now prove the theorem. Since A is real symmetric, it has three real latent roots (not necessarily distinct). Let λ_1 be such a root, then the equation

$$(\lambda_1 I - A)x = 0 \qquad (16)$$

has a non-zero solution $x=u$; if we normalize this solution it still satisfies (16) and we thus obtain a unit-vector u satisfying (16). By the lemma we can find a proper orthogonal matrix R with u as first column. By (16), $Au=\lambda_1 u$, and hence the matrix $AR=(Au, Av, Aw)$ has as its first column $\lambda_1 u$. Since R is orthogonal we find

$$R'AR=\begin{pmatrix} \lambda_1 & b_{12} & b_{13} \\ 0 & & \\ 0 & \multicolumn{2}{c}{B_1} \end{pmatrix},$$

where B_1 is a 2×2 matrix. Now A is symmetric, hence so is $B=R'AR$ and it follows that $b_{12}=b_{13}=0$; we now have

$$B=R'AR=\begin{pmatrix} \lambda_1 & 0 & 0 \\ 0 & & \\ 0 & \multicolumn{2}{c}{B_1} \end{pmatrix} \tag{17}$$

with a symmetric 2×2 matrix B_1. As before we can now find a latent root of B_1, a corresponding 2-dimensional unit-vector u_1 and using this as the first column of a proper orthogonal 2×2 matrix S_1 we obtain

$$S_1'B_1S_1=\begin{pmatrix} \lambda_2 & 0 \\ 0 & \lambda_3 \end{pmatrix}. \tag{18}$$

Now put

$$S=\begin{pmatrix} 1 & 0 & 0 \\ 0 & & \\ 0 & \multicolumn{2}{c}{S_1} \end{pmatrix},$$

then S is again proper orthogonal, and hence so is $P=RS$. Combining (17) and (18), we find

$$P'AP=S'R'ARS=S'BS=\begin{pmatrix} \lambda_1 & 0 & 0 \\ 0 & & \\ 0 & \multicolumn{2}{c}{S_1'B_1S_1} \end{pmatrix}=\begin{pmatrix} \lambda_1 & 0 & 0 \\ 0 & \lambda_2 & 0 \\ 0 & 0 & \lambda_3 \end{pmatrix}.$$

Thus A has been transformed to diagonal form by a proper orthogonal matrix and the proof is complete.

The algebraic content of the theorem just proved is that for any real symmetric matrix A, a proper orthogonal matrix P can be found such that $P'AP$ is diagonal. This holds quite generally, not only when A is 3×3, and the same method of proof can be used in the general case.

5. If u is any non-zero vector satisfying

$$(\lambda_1 I - A)u = 0,$$

then $|\lambda_1 I - A| = 0$, so that λ_1 is a latent root of A; u is then called a *characteristic vector* (or *eigenvector*) belonging to this root. It is useful to note that characteristic vectors belonging to different latent roots are orthogonal.

For, let $Au = \lambda_1 u$, $Av = \lambda_2 v$, then $v'Au = \lambda_1 v'u$, $u'Av = \lambda_2 u'v$, hence $(\lambda_2 - \lambda_1)u'v = u'Av - v'Au = 0$, and since we assumed $\lambda_1 \neq \lambda_2$, $u'v = 0$.

The preceding remark leads to a very short proof of the theorem of No. 4 in the case where A has distinct latent roots. We simply take a characteristic vector for each root. The three vectors are mutually orthogonal and normalizing them we obtain an orthogonal matrix $P = (u, v, w)$ such that

$$Au = \lambda_1 u, \; Av = \lambda_2 v, \; Aw = \lambda_3 w.$$

By changing the sign of w if necessary we can ensure that P is proper orthogonal. Now $AP = (Au, Av, Aw) = (\lambda_1 u, \lambda_2 v, \lambda_3 w)$ and by the orthogonality of P,

$$P'AP = \begin{pmatrix} \lambda_1 & 0 & 0 \\ 0 & \lambda_2 & 0 \\ 0 & 0 & \lambda_3 \end{pmatrix},$$

as we wished to show.

6. In practice it is often enough to know the latent roots, without having to compute the actual matrix of transformation to diagonal form. But even when this matrix is required, we usually find the latent roots first; the columns of the transforming matrix are then obtained by finding a characteristic vector for each root, taking care to choose the vectors belonging to a repeated root to be orthogonal. The process is illustrated in the following examples.

(i) $4x_1^2 + 2x_2^2 + 3x_3^2 - 4x_1x_3 - 4x_2x_3 + 3 = 0.$

The matrix of the quadric is

$$A=\begin{pmatrix} 4 & 0 & -2 \\ 0 & 2 & -2 \\ -2 & -2 & 3 \end{pmatrix}.$$

The characteristic equation is

$$\begin{vmatrix} \lambda-4 & 0 & 2 \\ 0 & \lambda-2 & 2 \\ 2 & 2 & \lambda-3 \end{vmatrix}=0,$$

which is $\lambda(\lambda^2-9\lambda+18)=0$. Hence the latent roots are 0, 3, 6.

The equation of the quadric in the new coordinates is

$$3y_2^2+6y_3^2-3=0, \text{ or } y_2^2+2y_3^2-1=0.$$

Each plane parallel to the 23-plane in the y-system cuts the surface in the same ellipse. It is therefore known as an *elliptic cylinder*.† We note that this is a non-central quadric, even though its equation can be expressed in the form (11).

To obtain the transforming matrix P we have to find the characteristic vectors.

Root $\lambda=0$. The characteristic vector $\boldsymbol{u}$ is obtained by solving

$$\begin{pmatrix} 4 & 0 & -2 \\ 0 & 2 & -2 \\ -2 & -2 & 3 \end{pmatrix}\begin{pmatrix} u_1 \\ u_2 \\ u_3 \end{pmatrix}=0.$$

A normalized solution is $u=\frac{1}{3}(1, 2, -2)'$.

Root $\lambda=3$. We have to solve the system

$$\begin{pmatrix} 1 & 0 & -2 \\ 0 & -1 & -2 \\ -2 & -2 & 0 \end{pmatrix}\begin{pmatrix} v_1 \\ v_2 \\ v_3 \end{pmatrix}=0.$$

A normalized solution is $v=\frac{1}{3}(2, -2, 1)'$.

Root $\lambda=6$. The system is

$$\begin{pmatrix} -2 & 0 & -2 \\ 0 & -4 & -2 \\ -2 & -2 & -3 \end{pmatrix}\begin{pmatrix} w_1 \\ w_2 \\ w_3 \end{pmatrix}=0.$$

† A quadric such as a cylinder, which can be written in the form (11) even though its matrix is singular, is sometimes said to have a line of centres (or in degenerate cases, a plane of centres).

A normalized solution is $w=\frac{1}{3}(2, 1, -2)'$.

It is easily seen that (u, v, w) is proper (e.g. by checking that $u \wedge v=w$). Hence the transforming matrix is

$$P=\frac{1}{3}\begin{pmatrix} 2 & 1 & 2 \\ 1 & 2 & -2 \\ -2 & 2 & 1 \end{pmatrix}.$$

(ii) $x_1^2+x_2^2+x_3^2+x_1x_2-x_1x_3-x_2x_3-1=0.$

If we multiply the equation by 2 (to avoid fractions) we obtain a quadric with characteristic equation

$$\begin{vmatrix} \lambda-2 & -1 & 1 \\ -1 & \lambda-2 & 1 \\ 1 & 1 & \lambda-2 \end{vmatrix}=(\lambda-2)^3-3(\lambda-2)-2=0.$$

The equation $x^3-3x-2=0$ has the roots $-1, -1, 2$, hence the latent roots are 1, 1, 4. The equation of the quadric in the new coordinates is therefore

$$4y_1^2+y_2^2+y_3^2-2=0.$$

The sections parallel to the 23-plane in the y-system are circles, other sections are ellipses. Thus we have an ellipsoid of revolution.

To find the transforming matrix we determine the characteristic vectors.

Root $\lambda=4$.

$$\begin{pmatrix} 2 & -1 & 1 \\ -1 & 2 & 1 \\ 1 & 1 & 2 \end{pmatrix}\begin{pmatrix} u_1 \\ u_2 \\ u_3 \end{pmatrix}=0.$$

Normalized solution $u=\frac{1}{\sqrt{3}}(1, 1, -1)'$.

Root $\lambda=1$.

$$\begin{pmatrix} -1 & -1 & 1 \\ -1 & -1 & 1 \\ 1 & 1 & -1 \end{pmatrix}\begin{pmatrix} v_1 \\ v_2 \\ v_3 \end{pmatrix}=0.$$

This system has rank one; in fact any vector orthogonal to u is a solution. We may take e.g. $v=\frac{1}{\sqrt{2}}(1, -1, 0)'$, and $w=u\wedge v=-\frac{1}{\sqrt{6}}(1, 1, 2)'$. The transforming matrix is therefore

$$P=\begin{pmatrix} \frac{1}{\sqrt{3}} & \frac{1}{\sqrt{2}} & -\frac{1}{\sqrt{6}} \\ \frac{1}{\sqrt{3}} & -\frac{1}{\sqrt{2}} & -\frac{1}{\sqrt{6}} \\ \frac{1}{-\sqrt{3}} & 0 & -\sqrt{\frac{2}{3}} \end{pmatrix}.$$

The freedom in the choice of v means that the directions of the characteristic vectors are not uniquely determined; this corresponds to the fact that we are dealing with a surface of revolution.

7. We have seen that every central quadric may, in a suitable coordinate-system, be expressed in the form

$$\lambda_1x_1^2+\lambda_2x_2^2+\lambda_3x_3^2+c=0. \qquad (19)$$

With the help of this form we can classify the different central quadrics without difficulty.

Suppose first that $c\neq 0$, then dividing (19) by $-c$ and taking the constant term to the right-hand side we may write the quadric as

$$\alpha_1x_1^2+\alpha_2x_2^2+\alpha_3x_3^2=1. \qquad (20)$$

(i) $\alpha_1, \alpha_2, \alpha_3$ all positive. Put $a_i=\alpha_i^{-\frac{1}{2}}$ $(i=1, 2, 3)$ then (20) represents an *ellipsoid* with semi-axes a_1, a_2, a_3 (Fig. 11). If two latent roots are equal, say $\alpha_1=\alpha_2$, then we have a *prolate* (elongated) or *oblate* (flattened) ellipsoid of revolu-

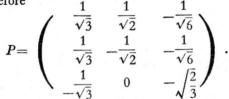

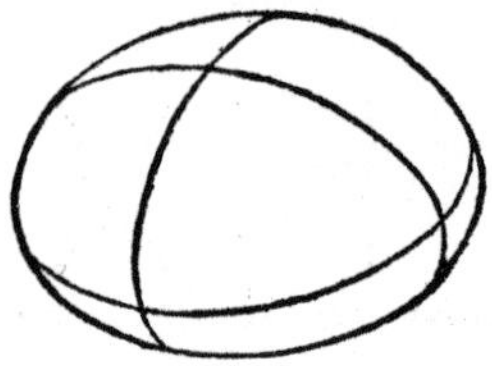

Fig. 11

tion according as $\alpha_3 < \alpha_1$ or $\alpha_3 > \alpha_1$. If all three roots are equal, we have a sphere.

(ii) One of $\alpha_1, \alpha_2, \alpha_3$ is negative, say $\alpha_3 < 0 < \alpha_1, \alpha_2$. The quadric meets the 1- and 2-axes, but not the 3-axis. It consists of a single piece which is cut by planes parallel to the 12-plane in an ellipse, and by planes parallel to the 13-plane or the 23-plane in a hyperbola. The quadric is called a *hyperboloid of one sheet* (Fig. 12).

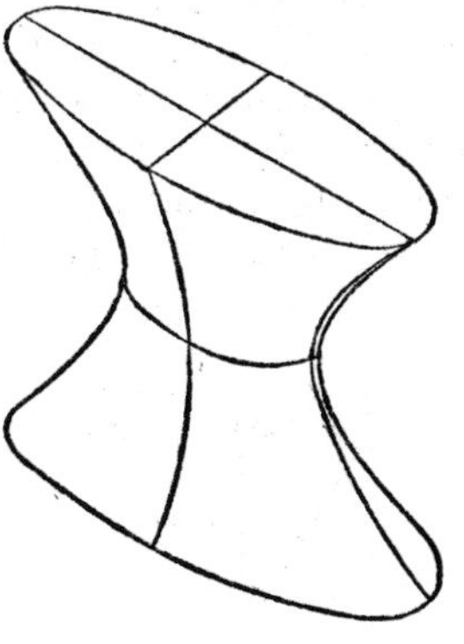

FIG. 12

(iii) Two of $\alpha_1, \alpha_2, \alpha_3$ are negative, say $\alpha_1 > 0 > \alpha_2, \alpha_3$. This is a *hyperboloid of two sheets* (Fig. 13). It consists of two

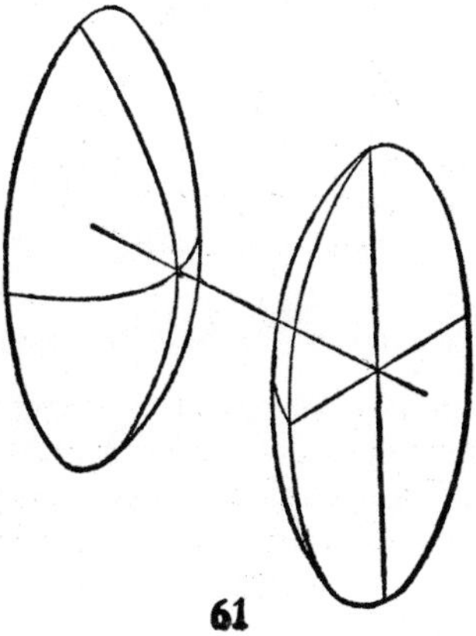

FIG. 13

separate pieces and meets the 1-axis, but not the 2- or 3-axis. A plane parallel to the 12-plane or the 13-plane cuts the surface in a hyperbola while a plane parallel to the 23-plane cuts it in an ellipse or not at all.

No other cases are possible, since in a central quadric $\alpha_1, \alpha_2, \alpha_3 \neq 0$ and in the remaining case where $\alpha_1, \alpha_2, \alpha_3$ are all negative, the equation (20) is not satisfied by any point.

If in (19) $c=0$, then either $\lambda_1, \lambda_2, \lambda_3$ are all of the same sign, in which case the origin is the only point on the surface, or two are of the same sign and one of the other. By permuting the coordinate-axes and if necessary multiplying the equation by -1, we may suppose that $\lambda_1, \lambda_2>0$ and $\lambda_3<0$. The resulting quadric is then a *cone* whose vertex (= centre) is at the origin (Fig. 14).

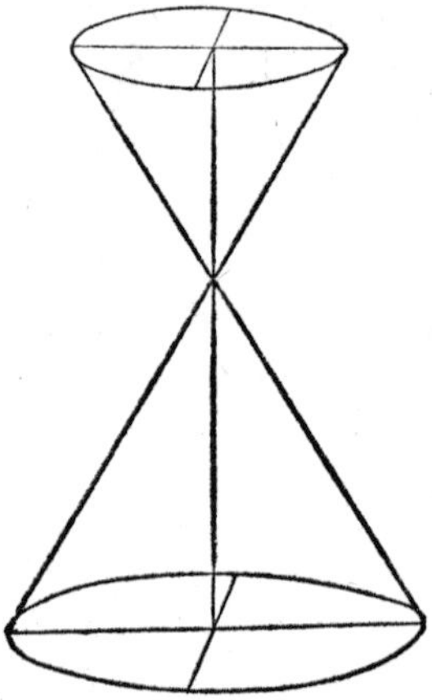

FIG. 14

8. It is useful to have a means of recognizing the type of a quadric from its equation without reducing this to the special form (19). The problem is essentially one of determining the signs of the latent roots of a symmetric matrix.

We shall provide an answer to this question in the next No. and at present confine ourselves to the case where all latent roots are positive.

A symmetric matrix A whose latent roots are all positive is called *positive-definite*. An equivalent definition is as follows: A is positive-definite if and only if

$$x'Ax>0 \tag{21}$$

for every non-zero vector x.

For, as we know,

$$x'Ax=\lambda_1y_1^2+\lambda_2y_2^2+\lambda_3y_3^2, \tag{22}$$

where $x=Py$ is the transformation which takes A to diagonal form. If all the latent roots λ_1, λ_2, λ_3 are positive, then the right-hand side of (22) is positive for all $y\neq 0$; but as y runs over all non-zero vectors, so does $x(=Py)$, hence (21) holds for all $x\neq 0$. Conversely, if (21) holds for all $x\neq 0$, then the right-hand side of (22) will be positive for all $y\neq 0$, and this is possible only if λ_1, λ_2, λ_3 are all positive.

Let A be given by

$$A=\begin{pmatrix} a_{11} & a_{12} & a_{13} \\ a_{21} & a_{22} & a_{23} \\ a_{31} & a_{32} & a_{33} \end{pmatrix} \quad (a_{ij}=a_{ji}), \tag{23}$$

then the numbers

$$d_0=1,\ d_1=a_{11},\ d_2=\begin{vmatrix} a_{11} & a_{12} \\ a_{21} & a_{22} \end{vmatrix},\ d_3=|A|$$

are called the *principal minors* of the matrix A. We note that each d_i (apart from the first and last) is the minor of the element in the bottom right-hand corner of d_{i+1}. Further, d_3, being equal to $|A|$, is just the product of the latent roots of A.

We have the following criterion for positive-definiteness.

Theorem. *A symmetric* 3×3 *matrix is positive-definite if and only if all its principal minors*† *are positive.*

† Clearly d_0 may be ignored here, since it is always 1.

To prove this result, let the matrix be given by (23) and denote the minor of a_{ij} in A by α_{ij}. We note that $d_1=a_{11}$, $d_2=\alpha_{33}$; suppose now that d_1, d_2, d_3 are all positive. We have

$$\begin{aligned} x'Ax &= a_{11}x_1^2+a_{22}x_2^2+a_{33}x_3^2+2a_{12}x_1x_2+2a_{13}x_1x_3+2a_{23}x_2x_3 \\ &= 1/a_{11}(a_{11}x_1+a_{12}x_2+a_{13}x_3)^2+1/a_{11}(\alpha_{33}x_2^2+\alpha_{22}x_3^2+2\alpha_{23}x_2x_3) \\ &= 1/a_{11}(\sum_i a_{1i}x_i)^2+1/a_{11}\alpha_{33}(\alpha_{33}x_2+\alpha_{23}x_3)^2+\beta/\alpha_{33}\,.\,x_3^2, \qquad (24) \end{aligned}$$

where

$$\beta=(\alpha_{33}\alpha_{22}-\alpha_{23}^2)/a_{11}. \qquad (25)$$

The expression on the right of (24) may be written $x'S'BSx$, where†

$$B=\begin{pmatrix} \dfrac{1}{a_{11}} & 0 & 0 \\ 0 & \dfrac{1}{a_{11}\alpha_{33}} & 0 \\ 0 & 0 & \dfrac{\beta}{\alpha_{33}} \end{pmatrix},$$

$$S=\begin{pmatrix} a_{11} & a_{12} & a_{13} \\ 0 & \alpha_{33} & \alpha_{23} \\ 0 & 0 & 1 \end{pmatrix}.$$

Thus $A=S'BS$, and taking determinants in this equation, we see that

$$\beta=|A|. \qquad (26)$$

Putting $y=Sx$, we may therefore rewrite (24) as

$$x'Ax=1/d_1\,.\,y_1^2+1/d_1d_2\,.\,y_2^2+d_3/d_2\,.\,y_3^2. \qquad (27)$$

Since d_1, d_2, d_3 are all positive, the right-hand side of (27) is positive for all $y\neq 0$; therefore $x'Ax>0$ for all $x\neq 0$ and so A is positive-definite.

Conversely, if A is positive-definite, then all the latent roots of A are positive and so $d_3=|A|>0$. Next we have

$$a_{11}x_1^2+a_{22}x_2^2+2a_{12}x_1x_2=x'Ax. \qquad (28)$$

where $x=(x_1,\ x_2,\ 0)'$; therefore the two sides of (28) are

† S need not be orthogonal, so that the coordinate-system used in 27) is oblique in general.

positive for all x_1, x_2 not both zero, which shows that the 2×2 matrix

$$\begin{pmatrix} a_{11} & a_{12} \\ a_{21} & a_{22} \end{pmatrix}$$

is positive-definite. Hence its determinant, d_2, is positive, and similarly $d_1=a_{11}$ must be positive because $a_{11}x_1^2>0$ for all $x_1\neq 0$. This completes the proof.

The theorem just proved holds in fact for symmetric matrices of any order, with a similar proof (using a generalization of the expression (24), which is due to Jacobi).

As an example we take the form

$$9x_1^2+13x_2^2+7x_3^2+12x_1x_2-6x_1x_3+2x_2x_3.$$

Its principal minors are $d_1=9$, $d_2=81$, $d_3=2025$, hence the form is positive-definite. As a second example consider the form

$$9x_1^2+13x_2^2+7x_3^2+24x_1x_2+6x_1x_3+2x_2x_3.$$

Here $d_1=9$, $d_2=-27$, $d_3=-243$, therefore this form is not positive-definite.

9. With the help of the criterion of No. 8 we can now give an expression for the number of negative latent roots of a symmetric matrix.

Theorem. *Let A be a regular symmetric 3×3 matrix and $d_0=1$, d_1, d_2, d_3 its principal minors. Then the number of negative latent roots of A (multiple roots being counted as often as they occur) is equal to the number of changes of sign in the sequence d_0, d_1, d_2, d_3. Here, any zero occurring between two non-zero numbers is omitted; if two successive zeros occur between two numbers of the same sign, the zeros are counted with the opposite sign; if the zeros occur between two numbers of opposite signs, they are omitted.*†

Let us denote by ν the number of changes of sign, as defined in the theorem. If the minors are, for example,

† Since d_0 and d_3 are both different from zero, not more than two consecutive zeros can occur.

1, 3, 0, -4, then $\nu=2$; if they are 1, 0, 0, 2, then $\nu=2$, while for 1, 0, 0, -7 we have $\nu=1$.

To prove the theorem suppose first that $d_3>0$. Then the latent roots λ_1, λ_2, λ_3 satisfy $\lambda_1\lambda_2\lambda_3>0$, so that the number of negative latent roots is even. If there are no negative latent roots, then d_1 and d_2 are positive, by the result of No. 8 and the result follows because there are now no changes of sign. If there are negative latent roots, then d_1 and d_2 cannot both be positive, again by No. 8, and the number of changes of sign is then two, whatever the actual values of d_1 and d_2.

Next let $d_3<0$. Then $-A$ has the principal minors

$$d_0,\ -d_1,\ d_2,\ -d_3, \tag{29}$$

and the latent roots

$$-\lambda_1,\ -\lambda_2,\ -\lambda_3. \tag{30}$$

In particular, $-d_3>0$, so that the first part of the proof applies to $-A$. Either the minors (29) are all positive, and then the roots (30) are all positive, or there are two sign changes in (29) and two negative roots in (30). Accordingly, either d_0, d_1, d_2, d_3 are alternately positive and negative and λ_1, λ_2, λ_3 are all negative, or at least one of the inequalities

$$d_1\geqslant 0,\ d_2\leqslant 0$$

holds and A has one negative latent root. Clearly there are three sign changes in the former case and one in the latter. Since d_3 cannot vanish (by the regularity of A), all the possibilities are exhausted and the theorem is established.

This theorem can again be extended to regular symmetric matrices of any order, if we assume in addition that A has not more than two successive vanishing principal minors.†

† This result is due to Frobenius.

EXERCISES ON CHAPTER FIVE

1. Transform to diagonal form, by an orthogonal transformation:

   (i) $3x_1^2+6x_2^2-2x_3^2+4x_1x_2-12x_1x_3+6x_2x_3$,

   (ii) $3x_1^2+3x_2^2+6x_3^2+2x_1x_2-4x_1x_3-4x_2x_3$,

   (iii) $x_1x_2+x_1x_3+x_2x_3$.

2. Given a central quadric, if a line through the centre meets the quadric in P and P', show that the tangent planes in P and P' are parallel.

3. For each of the following quadrics, find the tangent planes at the points where the 1-axis meets the quadric:

   (i) $x_1^2+x_2^2+2x_1x_3+2x_2x_3=1$.

   (ii) $2x_1^2-x_2^2+2x_3^2+2x_1x_2-4x_1x_3+2x_2x_3=18$,

   (iii) $4x_1^2+5x_2^2+5x_3^2+6x_1x_2+6x_1x_3+4x_2x_3=1$.

4. Find the latent roots of the matrices corresponding to the forms on the left of Ex. 3 (i)–(iii).

5. Classify the following quadrics (using the result of V.9):

   (i) $2x_1^2+17x_2^2-12x_1x_2+4x_1x_3-10x_2x_3-1=0$,

   (ii) $x_1^2+13x_2^2+28x_3^2+4x_1x_2+10x_1x_3+26x_2x_3-67=0$,

   (iii) $x_1^2+6x_2^2-x_3^2-2x_1x_2+2x_1x_3-12x_2x_3+20x_2-6x_3=0$,

   (iv) $x_1^2+9x_2^2-3x_3^2-8x_1x_2-2x_1x_3+8x_2x_3-2x_1-6x_2$
   $+2x_3-6=0$,

   (v) $x_1^2+3x_2^2+5x_3^2-2x_1x_2-4x_1x_3+4x_2x_3-4x_1+8x_2$
   $+10x_3+8=0$.

6. Let A be a symmetric matrix with latent roots λ_i. Show that the latent roots of the matrix $A-\alpha I$, where α is any scalar, are $\lambda_i-\alpha$.

7. Let A be a 3×3 symmetric matrix, and for any α, not a latent root of A, denote by $\delta(\alpha)$ the number of sign changes in the sequence of principal minors of $A-\alpha I$. Using Ex. 6, show that the number of latent roots of A less than α is $\delta(\alpha)$. Deduce that for any α and β which are not latent roots of A and such that $\alpha<\beta$, the number of latent roots of A between α and β is $\delta(\beta)-\delta(\alpha)$.

8. With the notation of V.8, show that the principal minors of a 3×3 symmetric matrix A satisfy the relations

$$d_0d_3=d_1a_{22}-a_{12}^2,$$
$$d_1d_2=d_2\alpha_{22}-\alpha_{23}^2.$$

   Deduce that if $d_3\neq 0$, and exactly one of d_1, d_2 vanishes, then its neighbours in the sequence of principal minors have opposite signs.

9. Let A and B be two symmetric matrices and assume that A is positive-definite. Show that there is a non-singular matrix P such that $P'AP=I$, $P'BP=D$, where D is a diagonal matrix with the roots of $|\lambda A-B|=0$ as diagonal elements.
   [Hint: By V.4, there is an orthogonal matrix Q such that $Q'AQ=C$

is diagonal, with positive diagonal elements γ_1, γ_2, γ_3 say. Let R be the diagonal matrix with diagonal elements $\gamma_i^{-\frac{1}{2}} (i=1, 2, 3)$ and take an orthogonal matrix S such that $S'(R'Q'BQR)S$ is diagonal; then $P=QRS$.]

In a dynamical system with generalized coordinates $q_1, \ldots, q_n$, let the kinetic energy near an equilibrium point be $T=\Sigma a_{ij}\dot{q}_i\dot{q}_j$, and the potential energy $V=\Sigma b_{ij}q_iq_j$ (to the second order in the q's), where $a_{ij}=a_{ji}$ and $b_{ij}=b_{ji}$. Then (a_{ij}) is a positive-definite matrix, and by transforming (a_{ij}) and (b_{ij}) simultaneously to diagonal form, as in the above exercise, we obtain a particularly simple form for the equations of motion (normal modes of vibration).

10. Let A be any regular symmetric 3×3 matrix, then A may be written as

$$A=P'DP,$$

where D is a diagonal matrix and $P=(u, v, w)$ is an orthogonal matrix. Show that the principal minors of A are

$$1, u'Du, |D| \,.\, w'D^{-1}w, |A|.$$

Answers to the Exercises

Chapter I:

3. $(7, 3, 5)$.
4. $(4, 3, -1)7/13$, $7\sqrt{2/13}$.

Chapter II:

In the answers to 1, 2 and 6, the position-vector of the general point on the line is given; note that this vector may take different forms.

1. (i) $(1, 1, 1)+\lambda(1, 0, 0)$, $(1, 1, 1)+\lambda(0, 1, 0)$, $(1, 1, 1)+\lambda(0, 0, 1)$, (ii) $(1, 1, 1)+\lambda(0, 1, 1)$, $(1, 1, 1)+\lambda(1, 0, 1)$, $(1, 1, 1)+\lambda(1, 1, 0)$, (iii) $(1, 1, 0)+\lambda(0, -1, 1)$, $(1, 0, 1)+\lambda(-1, 1, 0)$, $(0, 1, 1)+\lambda(1, 0, -1)$.

2. (i) $(1, 1, 1)+\lambda(3, 1, 2)$, (ii) $(2, 0, 1)+\lambda(5, 4, 1)$, (iii) $(1, 4, 2)+\lambda(-1, 1, 1)$.

3. (i) and (ii) $30°$, (i) and (iii) $90°$, (ii) and (iii) $90°$.

4. $1/\sqrt{3}$ (angle $54° \, 44' \, 8''$), $1/3$ (angle $70° \, 31' \, 44''$).

5. (i) $3x_1-10x_2+6x_3=1$; (ii) $7x_1+13x_2+14x_3=75$; (iii) $53x_1+5x_2-3x_3=54$.

6. (i) and (ii) $(1, 2, 3)+\lambda(2, 0, -1)$, (i) and (iii) $(1, 2, 3)+\lambda(0, 3, 5)$; (ii) and (iii) $(1, 2, 3)+\lambda(1, -7, 6)$.

7. (i) $(1, 2, -2)$, $(-1, -2, 2)$; (ii) $\frac{1}{3}(-1, 2, 2)$, $\frac{1}{7}(6, -2, 3)$.

9. $\mathbf{a}+\mathbf{b}-2\mathbf{c}$, $\mathbf{b}+\mathbf{c}-2\mathbf{a}$, $\mathbf{c}+\mathbf{a}-2\mathbf{b}$, $\mathbf{a}+\mathbf{b}+\mathbf{c}$.

Chapter III:

1. $x_1=y_3$, $x_2=y_1$, $x_3=y_2$.

2. Reflexion in the plane perpendicular to $\mathbf{a}$.

12. AB and CD: $8/\sqrt{13}$, AC and BD: $\sqrt{5}$, AD and BC: $8/\sqrt{5}$.

Chapter IV:

1. (i) $(-3, 0, 4)$, 5; (ii) $\frac{1}{3}(5, 4, 6)$, 3.
2. (i) $|x|^2-6x_1-2x_2-4x_3+5=0$; (ii) $|x|^2-2x_1-2x_2-2x_3-14=0$.
3. $|x|^2+2x_1-2x_3-36=0$.
4. $(1, 3, 3)$ and $(9, 3, -5)$; $x_1-7x_3=-20$, $7x_1-x_3=68$.
5. $(\mathbf{x}-\mathbf{a}) . (\mathbf{x}-\mathbf{b})=0$.
6. 4, $|x|^2-2x_1-2x_2+1=0$.

Chapter V:

1. (i) $7y_1^2+7y_2^2-7y_3^2$, where $\sqrt{5}y_1=x_1+2x_2$, $\sqrt{70}y_2=-6x_1+3x_2+5x_3$, $\sqrt{14}y_3=2x_1-x_2+3x_3$; (ii) $2y_1^2+2y_2^2+8y_3^2$, where $\sqrt{3}y_1=x_1+x_2+x_3$, $\sqrt{2}y_2=x_1-x_2$, $\sqrt{6}y_3=x_1+x_2-2x_3$; (iii) $y_1^2-\frac{1}{2}y_2^2-\frac{1}{2}y_3^2$ with the same transformation as in (ii) (in each case the transformation can take different forms, although the diagonal form itself is uniquely determined, to within a permutation of the variables).

3. (i) $x_1+x_2=\pm 1$; (ii) $2x_1+x_2-2x_3=\pm 6$; (iii) $4x_1+3x_2+3x_3=\pm 2$.

4. (i) $1, -1, 2$; (ii) $1, -2, 4$; (iii) $1, 3, 10$.

5. (i) Two-sheeted hyperboloid; (ii) ellipsoid; (iii) one-sheeted hyperboloid; (iv) cone; (v) no surface.

Index